DIAMOND AND CARBON COMPOSITES AND NANOCOMPOSITES

Edited by **Mahmood Aliofkhazraei**

Diamond and Carbon Composites and Nanocomposites

http://dx.doi.org/10.5772/61410
Edited by Mahmood Aliofkhazraei

Contributors

Ishak Afşin Kariper, Lucie Bacakova, Antonin Broz, Jana Liskova, Stepan Potocky, Lubica Stankova, Alexander Kromka, Damian Przestacki, Parveen Saini, Sebastian Suarez, Leander Reinert, Frank Mücklich, R. Vladoiu, Corneliu Porosnicu, Aurelia Mandes, Ionut Jepu, Virginia Dinca, Aurelian Marcu, Mihail Lungu, Liga Avotina, Gabriel Prodan

CBS, Edition **2017**

Published by InTech

Janeza Trdine 9, 51000 Rijeka, Croatia

Publishing Process Manager Iva Lipovic

Technical Editor InTech DTPTeam

Cover Designer InTech Design Team

Additional hard copies can be obtained from orders@intechopen.com

Diamond and Carbon Composites and Nanocomposites, Edited by Mahmood Aliofkhazraei
p. cm.
Print ISBN 978-953-51-2453-5
Online ISBN 978-953-51-2454-2

INTECH

open science | open minds

Contents

Preface

During the past few years, scientists have achieved significant successes in the nanoscience and technology. Nanotechnology is a branch of science that deals with fine structures and materials with very small dimensions - less than 100 nm. Measurement unit of nanometer has been extracted from nano prefix, which is a Greek word, meaning extremely fine. One nano (10^{-9} m) is the length equivalent to 5 silicon atoms or 10 hydrogen atoms aligned side by side. Besides, the composite science and technology have also benefits from nanotechnology. Nanotechnology is a field of applied sciences that focuses on design, production, detection, and employing the nano-size materials, pieces, and equipment. Advances in nanotechnology lead to improvement of tools and equipment as well as their application in human life. "Nano science" is the study of the phenomena emerged by atomic or molecular materials with the size of several nanometers to less than 100 nm. In chemistry, this size involves the range of colloids, micelles, polymer molecules, and structures such as very large molecules or dense accumulation of the molecules. In electrical engineering physics, nano science is strongly related to quantum behavior or electron behavior in structures with nano sizes. In biology and biochemistry, interesting cellular components and molecular structures such as DNA, RNA, and intercellular components are considered as nanostructures.

This book collects new developments about diamond and carbon composites and nanocomposites and their use in nanotechnology. I appreciate all the contributors to this book and thank them for their high quality manuscripts. I wish open access publishing of this book helps all researchers to benefit from this collection.

Dr. Mahmood Aliofkhazraei
Tarbiat Modares University,
Tehran, Iran
www.aliofkhazraei.com

Chapter 1

Hardness of Thin Films and the Influential Factors

İshak Afşin Kariper

Additional information is available at the end of the chapter

http://dx.doi.org/10.5772/63302

Abstract

Hardness is one of the most significant mechanical characteristics of a material. Hard materials are known for their durability. Currently, diamond is the hardest substance known in the world. Researchers have substantially worked on the production of this expensive material. Coating this material as a thin film was also the topic of a separate research. At the same time, hardness measurement, including the measurement of the hardness of a thin film was a topic of research as well. In this section, we will examine what researchers are doing for the measurement of the hardness, especially for bulk materials.

Keywords: Nanohardness, thin films, hardness, influential factors, diamond

1. Introduction

Hard materials are very useful for the industry and technology researches. SmCo5 and Nd2Fe14B can be used to produce televisions, video-recording devices, and speakers, with their magnetic-hard properties. Even consumers need and demand hard materials because they are long-lasting. Hard materials are highly effective when they are used to cover a surface. They are also used as diffusion barriers in electronic industry, as they preserve their chemical stability under high temperature and prevent the diffusion of foreign atoms. This allows electronic materials to be more durable than in the past (Mechanical corrosion is another important problem). Especially in mechanical industry, these materials (hard thin films) are very useful to prevent substances' mechanical corrosion. Based on the aforementioned reasons, nowadays, hard thin films have a significant place in coating.

In the abstract, we have underlined that hardness is one of the most important mechanical characteristics of a material. You can get insights about other characteristics of the material by measuring its hardness. Hardness is an indicator of the material's resistance against scratching, cutting, abrasion, and puncture. Table lists the hardness of some materials in terms of absolute hardness (**Table 1**).

Mineral	Absolute hardness
Talk: $Mg_3Si_4O_{10}(OH)_2$	1
Gypsum: $CaSO_4.2H_2O$	2
Calcite: $CaCO_3$	9
Fluorite: CaF_2	21
Quartz: SiO_2	100
Topaz: $Al_2SiO_4(OH,F)_2$	200
Ruby or Sapphire: $Cr–Al_2O_3$	400
Diamond: C	1500

Table 1. Absolute hardness of some materials.

Nanoscale, that is, a few hundred nanometers in size, materials are called nanomaterials. Thanks to nanomaterials, the technology has made great progress. The properties of a material in nanoscale are very different than the ones in the bulk form, which created great advantages for the technology. Especially, coating these materials to the surface as a thin film has added new properties to the base material. However, there is very little information in the literature on the mechanical properties of thin films [1].

After coating process, the coated material can gain electrical, optical, corrosive, and cracking resistance, depending on the thickness of the film. A material with such properties will be very handy in space research, astronomy, automobile industry, and in many other areas of engineering. For example, imagine that you can cover any surface with a heat-resistant material. This can be used in a wide range of area, from space engineering to the clothing of firefighters. On the other hand, if we could cover the surface of a material with diamond, we could make many materials harder and at the same time more robust [2].

When you produce such materials, another problem that you will be faced will be how to measure the hardness of the material when it is coated with such a thin film. In the literature, the hardness of thin films is generally measured by the indentation method [3], even though it has some limitations. The hardness of the substrate also affects the measurement, for example, while measuring, the hardness of soft thin films or a soft substrate may affect the hardness of the final product even though the coating that you have produced is very hard. Thus, the type of coating and the substrate is crucial.

Researchers are currently working on the measurement of the hardness of the thin films. All of them apply 10% rule in their studies. Accordingly, when a hollow-shaped hole is made on the surface of the film, the depth of this hole should be less than 10% of the film thickness. In this way, the effect of the base material to the hardness (and to the measurement) will be lower than 2% [4–8]. This is quite logical for the thin films, since their thickness is lower than 1000 nm; the depth of the hole should be around 100 nm. These units can be easily measured with the current technology [9]. In this topic, the impact of the hole's geometric shape on the mechanical properties, that is, hardness, also attracted the attention of the researchers [3]. We can see some outstanding studies in the literature about hardness of thin films in **Table 2**.

Author/s	Article	Material	Subject
M. Draissia, H. Boudemagh, and M. Y. Debili	Structure and hardness of the sputtered Al–Cu thin films system	Al–Cu thin film	Hardness of film according to % Cu [10]
• Lin-Dong Wang, • Min Li, • Tai-Hua Zhang, • Nai-Gang Liang	Hardness measurement and evaluation of thin film on material surface	None	Geometric effect [3]
• Seung Min Han, • Ranjana Saha, • William D. Nix	Determining hardness of thin films in elastically mismatched film-on-substrate systems using nanoindentation	Al/Si, Al/ sapphire, W/Si, and W/ glass	Hardness measurement [11]
Yu. A. Bykov, S. D. Karpukhin, Y. V. Panfilov, M. K. Boichenko, V. O. Cheptsov, and A. V. Osipov	Measurement of the hardness of thin films	TiN	Measurement of hardness and film thickness effect [12]
J. L. He, W. Z. Li and H. D. Li	Hardness measurement of thin films: separation from composite hardness	TiN, Cu	Measurement of hardness [13]
S. Chen, L. Liu, T. Wang	Investigation of the mechanical properties of thin films by nanoindentation, considering the effects of thickness and different coating–substrate combinations	Al, W	Measurement of hardness and film thickness effect [14]
Y.-G. Jung, B. R. Lawn, M. Martyniuk, H. Huang, X.Z. Hu	Evaluation of elastic modulus and hardness of thin films by nanoindentation	None	Elastic modulus and hardness of thin films [15]
S. Liu, H. Huang, Y. T. Gu	Hardness of silicon nitride thin films characterized by nanoindentation and nanoscratch deconvolution methods	SiN	Hardness of silicon nitride thin films [16]
N. Demas, C. Lorenzo-Martin, O.O. Ajayi, R.A. Erck, and I. Shareef	Measurement of thin-film coating hardness in the presence of contamination and roughness: implications for tribology	TiAlN, CrN	Hardness of thin film effect on tribology [17]

Author/s	Article	Material	Subject
P. D. Tall, S. Ndiaye, A. C. Beye, Z. Zong, W. O. Soboyejo, H.-J. Lee, A. G. Ramirez, K. Rajan	Nanoindentation of Ni–Ti thin films	Ni–Ti	Hardness of alloys thin film [18]

Table 2. Hardness of thin films in the literature.

In this chapter, we tried to formulize and explain the methods that can be used to measure the hardness of a thin film, the parameters affecting the hardness of a thin film, and the impacts of other parameters such as temperature, elasticity module, which also affect hardness, on the thin films.

2. Methods

The logic of hardness measurement is to indent the material to be tested with another material, whose hardness is known, and measure the resistance that the tested material displayed against it. The material that is used to measure the hardness (indenter) should be much harder than the tested material. Thus, mechanical defects that may occur during the measurement would be equilibrated. The geometric shape of the indenter is also important. It is usually chosen among the geometric shapes whose area can easily be calculated, such as ball (sphere), pyramid, or cone.

The usual method to achieve the hardness is to form a scratch, having an area (A), using an indenter by applying a specific force (F). Therefore, hardness H is

$$H = \frac{F}{A} \tag{1}$$

The area (A) formed on the surface of the material is negatively proportional with the hardness of the material. The following precautions should be taken while making the measurements:

1. The measurement should be made on a flat surface.
2. The thickness of the test material should be at least 10 times bigger than the depth of the indent (Therefore, microhardness measurements are not very suitable for thin films).
3. Hardness measurement should be made at the center of the sample.

   Currently, four methods can be mentioned for hardness measurement, as described in the following sections.

2.1. Brinell's hardness test

Brinell's hardness test consists of indenting a sphere with a 10 mm diameter into the material (**Figure 1**).

http://dx.doi.org/10.5772/63302

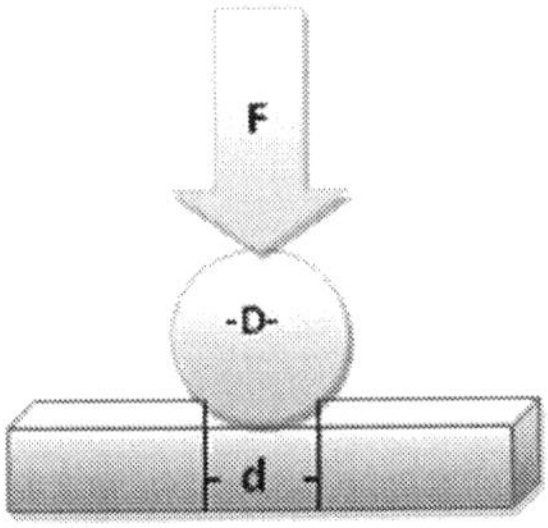

Figure 1. Brinell's hardness test.

Here, F is the applied force; D is the diameter of the indenter; and d is the diameter of the indentation.

$$H = \frac{2F}{\mu D(D - \sqrt{\square D^2 - d^2})} \tag{2}$$

The force to be applied is calculated according to $F = CD^2$, where C is the loading degree, varying for each material.

2.2. Rockwell's hardness test

It is similar to the Brinell hardness test; however, in this test, the holes between the surface of the sample and the indenter is equilibrated by applying a force of 10 kg. Then, an additional major load is applied and removed, while the preliminary load is still maintained (**Figure 2**).

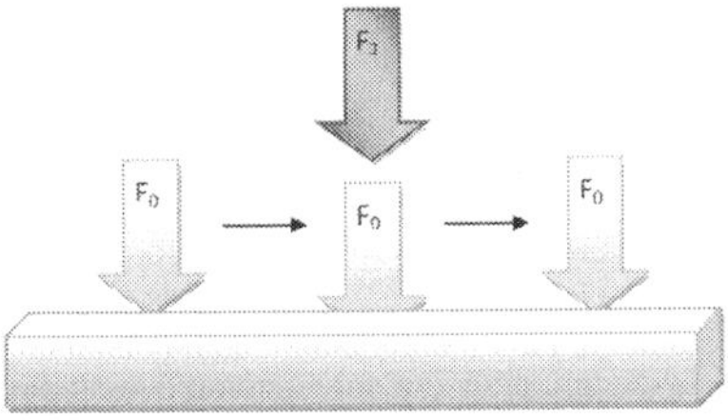

Figure 2. Rockwell's hardness test.

2.3. Vickers hardness test

In this test, force F is applied by a diamond indenter, in the form of a right pyramid with an angle of 136° between opposite faces (**Figure 3**). The calculations are made by measuring the diagonals of the indentation left on the surface of the material.

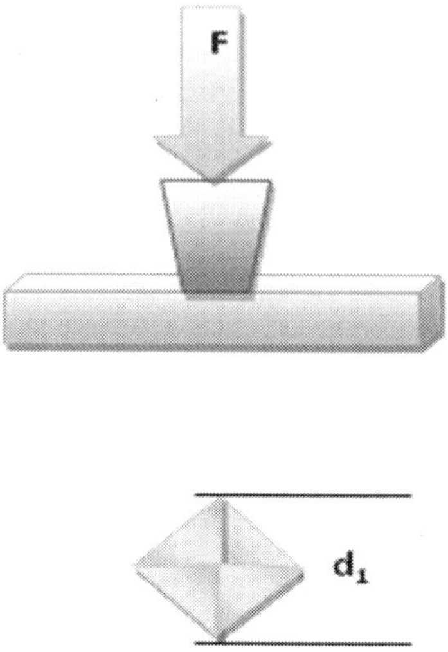

Figure 3. Vickers test and the indentation left in the surface of the material.

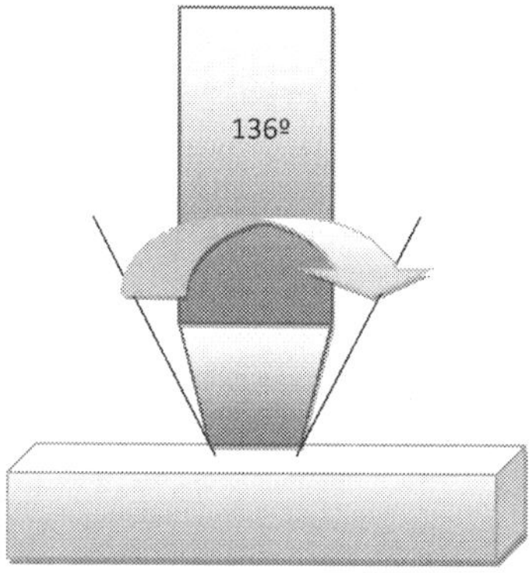

Figure 4. Indentation of the diamond into the material.

Using d_1 displayed above, d can be calculated as $d = (d_1 + d_2)/2$. And the hardness can be calculated as

$$H = 1.8544 \frac{F}{d^2} \quad (3)$$

Hardness of thin films can be measured with this method by using AFM devices.

2.4. Microhardness test

This method is suitable to measure the hardness of micron-sized samples. Similar to the Vickers test, a diamond indenter, in the form of a right pyramid, with an angle of 136° or 172°, is used (**Figure 4**). The hardness is

$$H = 14.2\frac{F}{d^2} \tag{4}$$

where d is the length of the indented edge [19].

3. Hardness of thin films

Researchers started to conduct researches on hard thin films and to produce them, because it became more and more important for the industry and the technology. Since the hardness measurement is as important as the thin films themselves, many researchers have focused on this issue as well. In various studies conducted until today, it has been indicated that the hardness of thin films can be calculated by indent images taken through nano-indentation device or AFM device (**Figure 5**).

In fact, these hardness measurements are also performed in a way similar to Vickers testing that we have explained above. The copper triple needle probe, which is used for tapping mode in AFM device, is indented to the surface by applying a particular force.

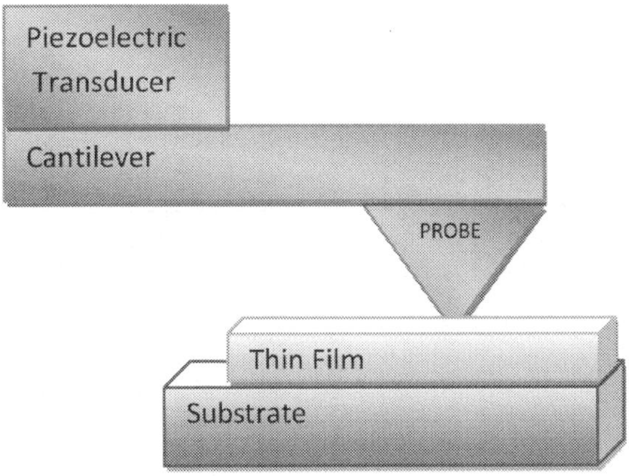

Figure 5. Hardness measurement by AFM.

The hardness of thin films can be determined using Vickers formulas (Eq. (3)), through the pyramid-shaped indentation seen on the surface image of the film. Here, measuring the hardness of the substrate along with the thin film is very important. Otherwise, you may end up measuring the hardness of the substrate, instead of the hardness of the nanometer-thick film. Therefore, it may be useful to compare the hardness of the substrate with the hardness of the film. To prevent this, researchers often use the following formulas:

$$E_c = \frac{\pi S}{2\beta\sqrt{A}} \tag{5}$$

where E_c: composite modulus,

S: measured stiffness,

B: Berkovich-type constant: 1.034.

Mechanical problems that may occur due to the substrate can be eliminated by shifting to the Young module E_m value:

$$E_m = \frac{\left(1 - v_m^2\right)}{\frac{1}{E^*} - \frac{1 - v_i^2}{E_i}} \tag{6}$$

where υ is the Poisson ratio and E is the Young module, representing the characteristics of the indenter and tested material with subindices "i" and "m". In this process, Young's module can be computed by replacing the indices with the tested film and the substrate, and the hardness of the film can be calculated.

Many researchers have worked on the hardness of thin films and described the parameters affecting the hardness of thin films. Some structural characteristics of thin films affect their optical, electrical, and mechanical properties. For example, it has been proven that electrical resistance of a thin film can be increased by decreasing its thickness; or film thickness and optic band gap are negatively proportional. However, researchers are a little late to investigate the impact of these parameters on the mechanical properties of thin films. On the other hand, there are researchers, such as Kariper, who have demonstrated that surface tension and surface energy of thin films are correlated with film thickness, average grain size, and number of crystalline per unit area.

According to literature, López et al. showed that film thickness and film hardness are negatively correlated [20]. Cavaleiro and Louro also observed that hardness is negatively correlated with grain size [21].

Due to the deformation of thin films, Ferro et al. could only show that there is a negative relation between thickness and hardness of the film: (i) the boundary between the film surface and the base prevents disordered structures and cracks, (ii) at the same time, nanocrystal nature of the films increases disordered structures. Therefore, it has been concluded that the decrease in the film thickness increases film tension of the area during the creation of the indent and causes the indent to be deeper; and this process prevents plastic deformation. These explanations have convergent parts with our interpretation [22].

Venkatraman and Bravman suggested that—based on the Hall–Petch effect—as the grain size decreases, structural dislocation decreases, and this process increases the flow stress [23].

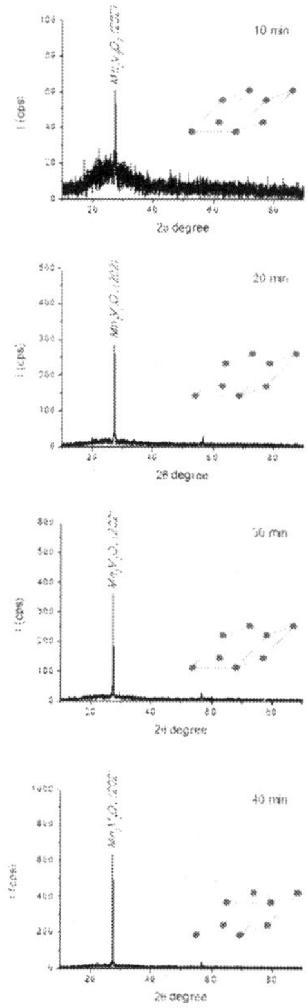

Figure 6. X-ray patterns of $Mn_2V_2O_7$ films with different deposition time: (a) 10 min, (b) 20 min, (c) 30 min, and (d) 40 min [24].

Kariper has published enormous and very clear works about the hardness of thin films and the influential factors [24]. He has demonstrated that parameters of thin films such as film thickness, grain size, surface roughness, etc. have affected the hardness of thin films. He had coated amorphous glass substrates with $Mn_2V_2O_7$ thin films. He showed that thickness and average grain size of the films varied with deposition time (The graphs and figures are taken from (and reorganized) the paper of Kariper published in the *International Journal of Minerals, Metallurgy and Materials*, in 2015). The lattice parameters of $Mn_2V_2O_7$ thin films were: $a = 0.6879$ nm, $b = 0.7973$ nm, $c = 1.0948$ nm, and the angles are $\alpha = 87.92°$, $\beta = 72°$, $\gamma = 83.29°$. The location of atoms in the plane is displayed in **Figure 6** and **Tables 3** and **Tables 4**. The peaks of the films were indexed with ASTM values (47-0342).

The researcher had measured the film thickness through AFM tapping mode; he calculated average grain size and number of crystalline per unit area through Scherrer's formulas; the obtained results are presented in **Figures 7** and **Figures 8**.

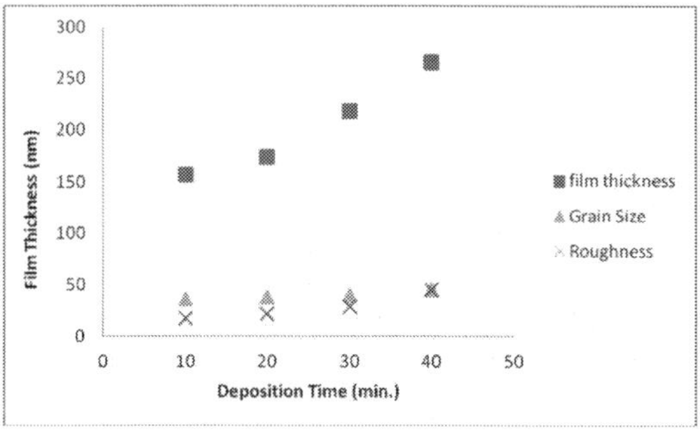

Figure 7. Film thickness, average grain size and roughness versus deposition time [24].

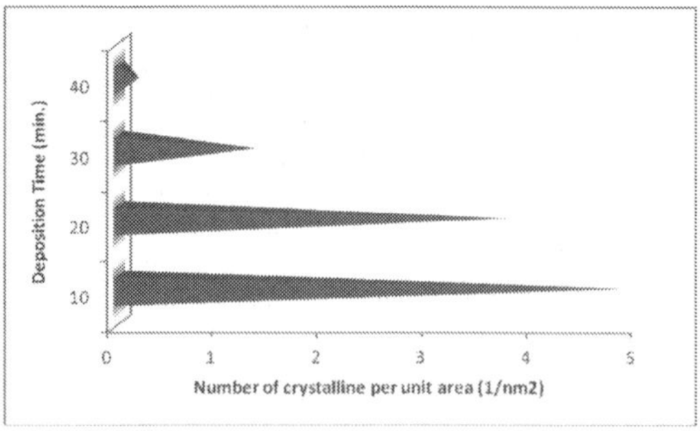

Figure 8. Number of crystalline per unit area and dislocation density of the films versus deposition time [24].

Deposition time (min.)	Film thickness (nm)	Grain size (nm)	Roughness (nm)
10	157	36	17
20	174	38	21
30	218	40	29
40	266	45	45

Table 3. Some parameters of the thin film according to deposition time [24].

Deposition time (min)	Number of crystalline per unit area (1/nm^2)
10	4,81
20	3,72
30	1,32
40	0,192

Table 4. Number of crystalline per unit area of the thin film according to deposition time [24].

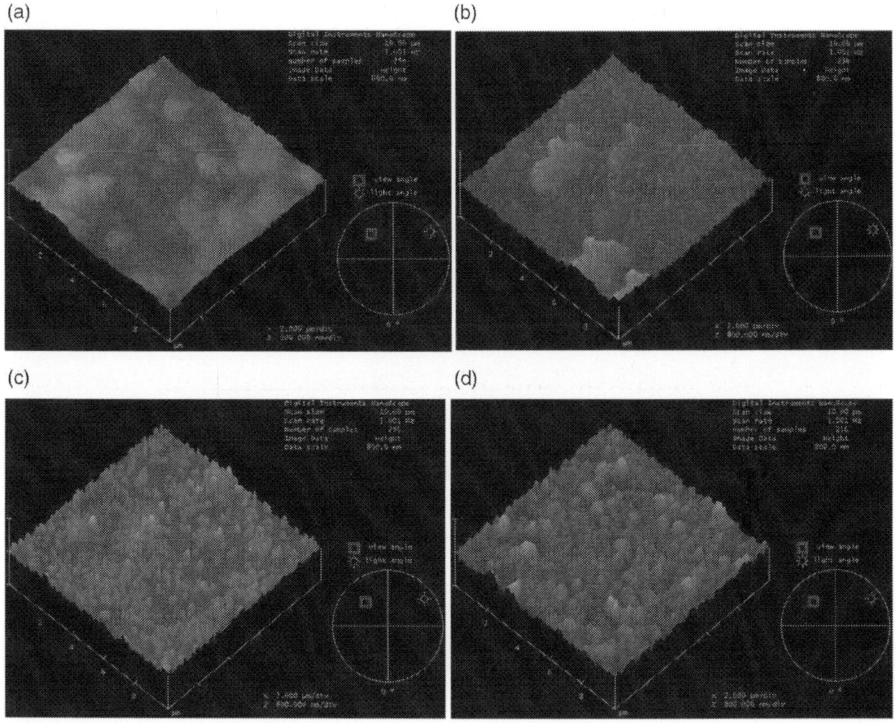

Figure 9. AFM images of the films with different deposition times (a) 10 min, (b) 20 min, (c) 30 min, and (d) 40 min [24].

He also stated that surface roughness affects hardness and published AFM surface images (**Figure 9**).

Afterward, he plotted the important parameters of thin films, such as film thickness, average grain size, surface roughness, and number of crystalline per unit area, in the form of a graph to find out if they are correlated with hardness (**Figure 10** and **Figure 11**, and **Table 5 (a)** and **(b)**).

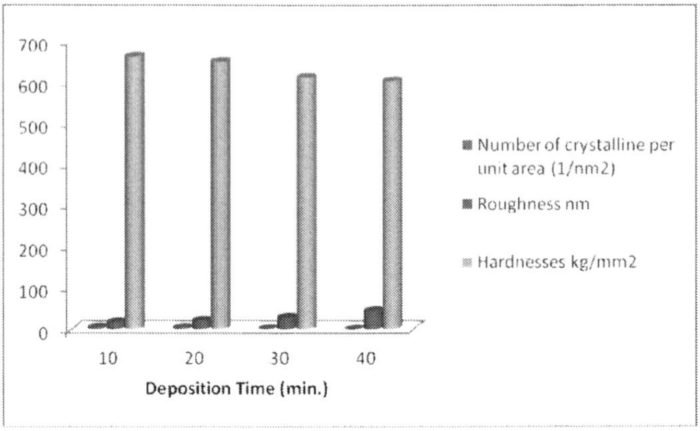

Figure 10. Relation between film thickness, average grain size, and hardness [24].

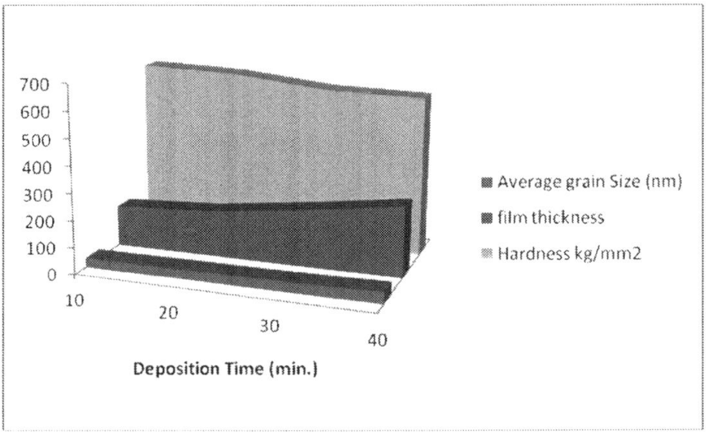

Figure 11. Relation between number of crystalline per unit area, roughness, and hardness [24].

Deposition time (min.)	Number of crystalline per unit area ($1/nm^2$)	Roughness (nm)	Hardness (kg/mm^2)
10	4,81	17	663
20	3,72	21	651
30	1,32	29	613
40	0,192	45	604

Table 5. (a) Some parameters of the thin film according to hardness [24].

http://dx.doi.org/10.5772/63302

Deposition time (min.)	Film thickness (nm)	Average grain size (nm)	Hardness (kg/mm²)
10	157	36	663
20	174	38	651
30	218	40	613
40	266	45	604

Table 5. (b) Some parameters of the thin film according to hardness [24].

He achieved important findings with the outcomes of these graphs:

H = -1.6343t + 1237.9 (H: hardness, t: film thickness)

H= -0.1222D + 117.09 (H: hardness, D: average grain size)

H= 0.0738N + 44.162 (H: hardness, N: number of crystalline per unit area)

H= -0.3931R + 276.76 (H: hardness, R: roughness)

In spite of some small experimental errors, he suggested that these formulas can be combined, and can be expressed as follows:

$H = 0.01845N - 0.4086t - 0.03055D - 0.0983R + 418.978$.

4. Elasticity module of thin films

Hooke is a British scholar who lived in London between the years 1635 and 1703, and he provided great convenience to the concept of strength with the law he has discovered. Hooke has formed his law for isotropic material as follows:

For instance, let us take two different steel bars and apply two different forces ($F1$ and $F2$) to these bars in the dynamometer. Steel bars will be affected by two different tensions, $\sigma 1$ and $\sigma 2$, where $\sigma_1 = F_1/A_1$ and $\sigma_2 = F_2/A_2$. We can find out the elongation of the bars, $\Delta L1$ and $\Delta L2$. The ratio of these elongations to $L01$ and $L02$ gives us the stretching. It can be find "elasticity module," E, which is one of the main characteristics of a material, by writing the formulas of these elongations and divide them to the tensions that we have calculated above. Thomas Young first found this value; thus, it is called as "Young's modulus" in the literature.

Here, we cannot measure the elongation of flexible thin films, because we have coated them on a substrate. However, we can get insights about their elasticity module and elasticity indirectly and make comparisons ($\sigma = E\varepsilon$, where E is elasticity module and ε is unit elongation).

$1/E\alpha$ - 1.6343t + 1237.9 (t: film thickness)

$1/E\alpha$ - 0.1222D + 117.09 (D: average grain size)

$1/E\alpha$ 0.0738N + 44.162 (N: number of crystalline per unit area)

$1/E\alpha$ -0.3931R + 276.76 (R: roughness)

Chapter 2

Graphene-Based Composites and Hybrids for Water Purification Applications

Rahul Sharma and Parveen Saini

Additional information is available at the end of the chapter

http://dx.doi.org/10.5772/63666

Abstract

The increasing water demand and the worldwide shortage of clean water call for new technologies for wastewater treatment, of which sorption is recognized as simplest and efficient method for all major water pollutants, including heavy metal ions, organic dyes and organic solvents. In this context, graphene, an atomically thin two-dimensional hexagonal arrays of carbon atoms, and its analogs have been visualized as ultimate materials for the separation of pollutants from water due to properties such as superhydrophobicity, superoleophilicity, chemical-environmental-thermal stability, chemical manipulability, controllable morphology, texture, porosity and accessible surface area, facile processing, low density, biocompatibility and low-cost synthesis from a variety of abundant carbonaceous precursors. This chapter begins with an introduction of water purification, related processes and fundamentals of adsorption and absorption phenomenon. The next section emphasizes the importance and advantages of graphene as sorbent, brief about its synthesis, types of graphene-based composites/hybrid systems and their synthesis strategies. The last section throws light on the performance of graphene-based foams, aerogels and composites/hybrids for the separation of metal ions, organic dyes and various oils/organic solvents, supported by comprehensive literature account. Finally, the chapter concludes with the comments on the current status, major challenges and future scope in the direction.

Keywords: sorption, adsorption, absorption, graphene oxide (GO), reduced graphene oxide (RGO), superhydrophobicity, superoleophilicity

1. Introduction

In the current scenario of global socioeconomic development, environmental conservation is a major concern and numerous initiatives have been undertaken to handle air, water, soil, noise and electromagnetic pollutions [1–13]. In particular, water pollution is recognized as a big threat for society, and enormous efforts are underway to prevent contamination of freshwater and to purify contaminated water. The water pollution is considered as an offshoot of scientific and technological growth, industrial revolutionization and ignorance/unavailability of efficient waste management strategies. The discharge of domestic and industrial effluents/wastewater into water bodies not only makes water sources unfit for drinking purpose but also adversely affects its utility for pharmaceutical, agricultural, chemical and industrial applications. In particular, dumping of untreated wastewater at domestic and industrial levels (e.g., by paper, textile, food, leather and chemical industries) introduces hazardous substances, for example, dyes, chromium compounds, heavy metals such as copper, arsenic, lead, cadmium, mercury, nickel, cobalt and certain toxic organic chemicals, into rivers, lakes and other freshwater bodies [10,14–18]. These contaminants affect the photosynthesis in aquatic systems and disturb the self-purification mechanism [19] of water with extinction of some aquatic species as one of its worst consequence. Their acute exposure poses serious health hazards for human, for example, organic dyes and heavy metal ions are known to exert harmful effects such as an increased heart rate, vomiting, shock, cyanosis, jaundice, tissue necrosis, bone defects, renal damage, paralysis and even death [20–22] in some cases.

The leakage and spilling of industrial chemicals and oils into water bodies also disturb the water quality and adversely affect the aquatic life. The large-scale oil spill accidents (occurring during onshore/offshore handling of oils) are identified as catastrophic and devastating situation for marine life. It also adversely affects the infrastructure (e.g., beaches), underwater ecosystem (fauna and flora) and badly affects the economies in terms of losses in tourism, fishing, seafood sectors and expenditures for the restoration of the same [23–31]. Therefore, in the past half-century, an extensive research was carried out across the globe to develop a number of efficient and smart materials/technologies for the effective treatment of wastewater containing various pollutant, especially metal ions, dyes, oils and solvents [32–42,17,43–45]. Among currently available mechanical, chemical and biological water purification processes such as coagulation, flocculation, ozone treatment, chemical oxidation, membrane filtration, ion exchange, photocatalysis and sorption [46–56], sorption (adsorption-absorption) practices emerged as most preferred commercially viable solution, owing to specific advantages in terms of simplicity of process, cost-effectiveness, environmental friendliness, applicability to diversified pollutants and low-energy consumption. In particular, adsorption practices are the sole most promising alternates in the isolation of metal ions that are otherwise very difficult to remove by means of photocatalysis and membrane filtration [57,58]. However, adsorption has also been demonstrated as viable solution for the removal of hazardous organic dyes from the wastewater streams of various industries. In contrast, absorption practices are prevalent in selective removal of nonpolar (hydrophobic) pollutants such as mineral or vegetable oils, crude oils and common organic solvents [28,59,60] from their aqueous emulsion or floating layers over water.

In the past, a number of adsorbing-absorbing materials have been exploited for wastewater treatment, including zeolite [61], silica gel [62], activated alumina [63], fly ash [64], polymeric and carbon-based materials [65–68]. Among them, carbon-based systems with environmental

http://dx.doi.org/10.5772/63666

compatibility, large-scale availability, high porosity, large specific surface area and low cost are the most promising. Activated carbon derivatives with high degree of porosity and large specific surface area (600–2000) m^2g^{-1} are used in adsorption-based water purification for many years [65,69–72]. However, essentiality of their surface activation to enhance adsorption capacity and release of secondary pollutants in the form of fine particles in water limited their utility. The new members of the carbon family fullerene, carbon nanotube (CNT) and graphene have truly revolutionalized the field with graphene, the latest member of the family, seems to be most promising alternate owing to its extraordinary mechanical, thermal & chemical properties coupled with very high specific surface area, inherent hydrophobicity, tunable chemical structure, cost-effective and scalable synthesis [10,73–78]. In this context, the quest for the development of highly efficient adsorbents and absorbents and advancements in graphene science and technology fuelled numerous efforts for designing/selecting highly efficient graphene-based sorbents, understanding their pollutant-capturing mechanism and assessing their actual purification performance [16,47,57,79–83,43].

It is worth mentioning that though sorption is a broad term, the underlying adsorption and absorption subprocesses are surface- and bulk-related phenomenon, respectively, and their governing factors are entirely different. For example, adsorption process depends on parameters such as nature of adsorbent-adsorbate pair, accessible surface area of adsorbent, concentration of pollutants, temperature, pH, etc [15,45,48,84]. In contrast, absorption is regulated by pore size/volume of the sorbent, its texture and wettability characteristics along with physical attributes of absorbate species and absorbent-absorbate interfacial properties [75,85,86]. Therefore, detailed understanding of adsorption and absorption phenomenon, related terminology, important parameters and governing empirical equations is necessary before identifying and designing a suitable graphene-based adsorbent or absorbent.

2. Adsorption

The term adsorption implies the presence of an excess concentration of molecular species (gas or liquid), known as adsorbate (**Figure 1**), at the surface of porous solid (i.e., at solid-fluid interface), known as adsorbent, compared to its concentration in bulk of the adsorbent.

Adsorption arises due to the fact that the surface particles of the adsorbent are not in the same environment as the particles inside the bulk. In solids, unsatisfactory valence forces of surface

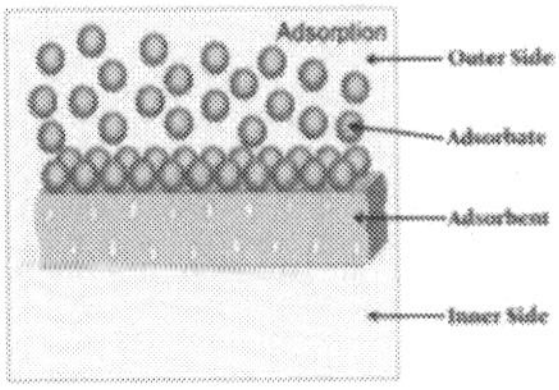

Figure 1. Pictorial representation of adsorption process.

atoms, activate the surface. At these activated surfaces, other substances stick feasibly with a decrease in the entropy (S) of the system. Since every spontaneous process is driven by a decrease in Gibbs free energy (G) that can be expressed as follows:

$$\Delta G = \Delta H - T\Delta S \tag{1}$$

Therefore, adsorption is essentially an exothermic process and the magnitude of change in enthalpy (ΔH) must be greater than the magnitude of term '$T\Delta S$'. As the adsorption proceeds, ΔH becomes less negative and ultimately becomes equal to $T\Delta S$, so that ΔG becomes zero and equilibrium is attained.

The term adsorption involves the capture of adsorbate exclusively over the surface of adsorbent. Therefore, on the basis of type interaction between adsorbent and adsorbate (i.e., physical or chemical), adsorption can be classified into two main categories, namely physisorption that involves van der Waals binding forces and chemisorption [87] wherein a strong chemical bond formation takes place between adsorbent and adsorbate. Adsorption is a complex phenomenon and requires systematic studies to understand completely its mechanism, kinetics and the thermodynamic parameters that influence adsorption capacity and rate. In particular, homogeneous and heterogeneous adsorption types have been identified with the involvement of single or multilayer adsorption of adsorbates [87]. To understand such processes and to account for the equilibrium adsorbate concentration over adsorbent and bulk of solution, several adsorption isotherms and related governing equations have been proposed, which are described in the following section.

2.1. Freundlich adsorption isotherm

Freundlich, in 1909, gave an empirical relationship between the quantity of a substance adsorbed by unit mass of solid adsorbent and the equilibrium concentration of adsorbate solution at a particular temperature for heterogeneous surfaces, which can be expressed by the following Eq. (2) [70]:

$$x / m = q_e = kC_e^{1/n} \tag{2}$$

where 'x' is amount (in mol) of adsorbate adsorbed by 'm' units amount (in g) of adsorbent. The constants K and n are the characteristics of given adsorbent-adsorbate pair. C_e is the equilibrium concentration (i.e., when adsorption is complete) of adsorbate solution. The use of logarithmic function converts Eq. (2) into the following linear form:

$$\log q_e = \log(x / m) = \log K + \frac{1}{n}\log C_e \tag{3}$$

Plot of log(x/m) versus log C_e gives a straight line and values of 'n' and 'K' can be calculated from its slope and intercept, respectively. The validity of Freundlich adsorption isotherm depends upon the completion of adsorption process. Equal volumes of standard solutions are added to equal amounts (m) of adsorbent (say charcoal) in different flasks. The final concentration (C_e) is determined in each flask after adsorption. The difference in the initial and final concentrations gives the value of 'x'. For each set of 'x' and 'C_e' values, the validity of Freundlich isotherm can be established using Eq. (3).

2.2. Langmuir adsorption isotherm

Langmuir derived a quantitative expression for variation in the extent of adsorption with concentration, by assuming the equilibrium nature of adsorption and fixed number of equivalent energy adsorbing sites on adsorbent surface [70] .This model is applicable to homogeneous surfaces and assumes that only monolayer of adsorbate molecules formed on adsorbent surface and there is no interaction between the adsorbate species after adsorption. This isotherm represents the equilibrium distribution of a material in solid and liquid or gas phases and can be mathematically expressed as follows [70]:

$$x / m = q_e = \frac{qKC_e}{1 + KC_e} \tag{4}$$

where q_e and C_e are the equilibrium adsorption capacity (mol/g) and equilibrium concentration of adsorbate solution, respectively. The constants K and q are the Langmuir's constant and maximum monolayer adsorption capacity. The linear form of Langmuir Eq. (4) is given as follows:

$$\frac{1}{q_e} = \frac{1}{q} + \frac{1}{qKC_e} \tag{5}$$

By plotting $1/q_e$ against $1/C_e$ value of K and q can be calculated from slope and intercept, respectively, of fitted straight line. In order to judge the feasibility of adsorption, a dimensionless constant parameter R_L called separation factor is defined, which is related to Langmuir constant and initial concentration C_i of adsorbate solution as follows [70]:

$$R_l = \frac{1}{1 + KC_i} \tag{6}$$

This constant explains the characteristic features of Langmuir's adsorption isotherm, as it indicates the shape of isotherm/energy of the adsorption. The magnitude of R_L signifies the feasibility of adsorption as $R_L>1$, $R_L=1$, $1<R_L>0$, $R_L=0$ indicates unfavorability, linearity, favorability and irreversibility of adsorption, respectively [70].

2.3. Temkin adsorption isotherm

According to Temkin adsorption isotherm, energy of the adsorption is a linear function of adsorbent surface coverage by assuming the indirect interaction of adsorbent and adsorbate [18]. The heat of adsorption of adsorbate molecules decreases with increase in the surface coverage of adsorbent [18]. This model can be represented by the following mathematical equations:

$$q_e = \frac{RT}{b} \ln kC_e \tag{7}$$

$$q_e = B \ln K + B \ln C_e \tag{8}$$

where q_e is the equilibrium adsorption capacity (mol g^{-1}) and C_e is the equilibrium concentration of adsorbate solution. Plots of q_e versus C_e give a straight line whose slope provides the values of *B*, which is associated with energy of adsorption, and Temkin's constant *K* can be calculated by intercept of the line.

A number of other adsorption isotherm models, namely BET isotherms [88], Dubinin-Radushkevich isotherm [89], Toth adsorption isotherms [90] are proposed to describe the adsorption phenomenon. Since precise mechanism of adsorption in solution phases is still not known to a greater extent; therefore, a single adsorption model is not applicable universally. The validity of these models generally changes from one adsorbent-adsorbate system to another and also with pH of solution, nature of solvent, ionic strength, etc. [15,91,18]. Freundlich and Langmuir adsorption isotherms are the two very common approaches generally used in the experimental practices. Since the use of adsorption isotherm equations essentially involves the 'curve-fitting' method, the model with higher value of regression coefficient is considered to be more valid than the one with lower regression coefficient value.

2.4. Thermodynamic treatment of adsorption isotherm

Since every spontaneous process proceeds by characteristic changes in thermodynamic parameters, thermodynamic parameters, such as Gibbs free energy change (ΔG_{ads}), change in enthalpy (ΔH_{ads}) and change in entropy (ΔS_{ads}) associated with the adsorption phenomenon, provide thorough knowledge about energy changes involved in adsorption process and the other thermodynamic variables on which adsorption depends on parameters such as temperature, pressure, etc. By designing systematic experiments, we can find out the distribution coefficient K_d. More specifically, it is equilibrium constant for the distribution of adsorbate molecules in solution phase and on adsorbent surface and given by Eq. (9):

$$K_D = \frac{C_{ea}}{C_e} \tag{9}$$

where C_{ea} and C_e are the equilibrium concentrations of adsorbate molecules on adsorbent surface and solution respectively. Other thermodynamic properties can be deduced using the distribution coefficient K_d.

$$\Delta G_{ads} = -RT \ln K_d \tag{10}$$

$$\ln K_d = \frac{\Delta S_{ads}}{R} - \frac{\Delta H_{ads}}{RT} \tag{11}$$

The values of the ΔH_{ads} and ΔS_{ads} will be easily determined using the van't Hoff curves by plotting K_d against T^{-1} which gives a straight line. The slope and intercept of line gives ΔS_{ads} and ΔH_{ads}, respectively. The decrease in the Gibbs free energy, that is, a negative value of ΔG, is the necessity of a spontaneous process.

2.5. Kinetics of adsorption

Thermodynamics deals with spontaneity of the processes, but it does not tell about rapidity (e.g., adsorption rates) of the adsorption process and its mechanistic details. Therefore, adsorption kinetic studies are employed for better understanding of adsorption mechanism and its rate. Since adsorption in solution phase involves more than two components viz. adsorbate, solvent and adsorbent, depending upon the adsorption system, several kinetic models have been proposed.

2.5.1. Pseudofirst-order kinetic model

In 1898, Lagergren proposed the pseudofirst-order kinetic model [15,92] for the adsorption of adsorbate particles from solution by a solid adsorbent and proposed the following governing mathematical Eq. (12):

$$\log(q_e - q_t) = \log q_e - \frac{K_1 t}{2.303} \tag{12}$$

where K_1 is the pseudofirst-order rate constant, q_e and q_t are equilibrium adsorption capacity (mol/maximum adsorption capacity for a given amount of adsorbent beyond which no further adsorption can take place) and adsorption capacity (mol/g) at a given time (t), respectively. By plotting $\log(q_e - q_t)$ versus 't' and linear data fitting values of K_1 and q_e from slope intercept, respectively. It has been realized that pseudofirst-order kinetic model is generally applicable to the initial stage of the adsorption as it only account for the involvement of only two components in adsorption process that occurs in solution phase [93,94].

2.5.2. Pseudosecond-order kinetic model

A more appropriate kinetic model for solution phase adsorption process is proposed by Ho and McKay [95] who showed that the rate of adsorption depends upon the adsorption capacity of adsorbate on solid adsorbent rather than its concentration in the solution phase. Eq. (13) was proposed to investigate the kinetics [96]:

$$\frac{1}{q_t} = \frac{1}{q_e K_2} + \frac{1}{q_e} \tag{13}$$

where K_2 and q_e represent pseudo-second-order rate constant and equilibrium adsorption capacity, respectively. Again the linear fitting of 't/q_t' versus 't' plots allowed the determination of K_2 and q_e from the calculated intercept and slope values.

For a better look into the mechanistic pathways, one should accounts the three basic steps by which adsorption proceeds viz. (i) the movement of adsorbate particles through bulk solution to the external surface of adsorbent, (ii) diffusion of adsorbate particles in the pores of adsorbent (interparticle diffusion) and (iii) adsorption of the adsorbate particles on the active sites of adsorbent. Any of the above steps can be rate-determining step.

For the identification of various steps involved in the adsorption process, interparticle diffusion model is used, which is expressed as follows [97]:

$$q_t = K_i T^{1/2} + C \tag{14}$$

where K_i is the interparticle diffusion rate constant and C is a numeric constant. If interparticle diffusion is the rate-determining step of the adsorption process, then the above equation gives straight line for q_t versus $t^{1/2}$ plots.

3. Absorption

Absorption is a phenomenon in which chemical substances, such as atoms, molecules and ions, enter into the bulk phase of absorbent and distributed throughout the bulk of absorbent. The process of absorption can be either physical [98] or chemical [98] depending upon the interaction between the absorbent and the absorbing species. Physical absorption is sometimes referred as a nonreactive process, for example, the absorption of air oxygen in water at air-water interface [99]. Such absorption phenomenon is highly temperature and pressure dependent. In practice, the absorbing capacity of the absorbent depends upon the interfacial surface area, contact time and the wettability characteristics of the both substances. For example, superhydrophilic materials preferentially absorb water molecules from a mixture of several substances, whereas superhydrophobic surfaces (e.g., lotus leaf) tend to exclude water. A reactive or chemical absorption process, which takes place by means of some chemical reactions, for example, the absorption of carbon dioxide by sodium hydroxide, occurs via the formation of sodium carbonate [99]. Since the absorption of adsorbate occurs through the surface of absorbent, surface properties of the solid absorbent are most critical in deciding their absorption capacity, absorption rate, etc. These properties include surface energy [100], surface wettability, [100] topological roughness [101] and contact angles [101]. For absolutely flat surfaces, surface wettability can be determined using Young's equation that correlates the interfacial surface energies with contact angle value by the following equation:

$$\cos\theta_r = \frac{\gamma_{sv} - \gamma_{sl}}{\gamma_{lv}} \tag{15}$$

where γ_{sv}, γ_{sl}, γ_{sl} are surface-free energy of involved solid-gas, solid-liquid and liquid-gas interfaces, respectively, and θ_r is the water contact angle with solid surface (**Figure 2a**). Therefore, for a solid absorbent, low surface energy is required for higher water contact angles (WCAs). Since lower contact angles are the direct reflection of surface wettability, for achieving highly wettable or superhydrophilic surfaces, we need to design the surfaces having minimum surface energies and an opposite situation is required for superhydrophobic surfaces having excellent wettability for nonpolar liquids such as oils and organic solvents. In addition to this, surface roughness and topology are another crucial factors for high/low WCAs, which are explained on the basis of Wenzel [102] (**Figure 2b**) and Cassie-Baxter [103] (**Figure 2c**) models.

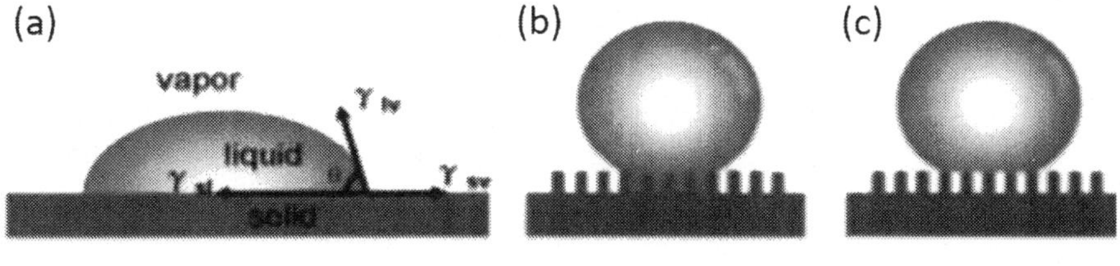

Figure 2. Schematic representation of wetting behavior of water droplet on solid absorbent (a) Young's model, (b) Wenzel model, (c) Cassie model. Reproduced from [75] with permission from RSC.

According to Wenzel's theory, no air bubbles are present between the liquid and solid surface. Hence, the liquid floods the roughness introduced channels (**Figure 2b**) leading to perfect wetting of the underlying surface, so that the contact angle of smooth (nontextured) and structured (rough) surfaces satisfy the following equation:

$$\cos\theta_\omega = \mathrm{r}\cos\theta \tag{16}$$

where θ_ω and θ are the contact angles of structured (rough surface) and smooth (nontextured) surfaces respectively, '*r*' is the roughness ratio (i.e., ratio of the true surface area of the structured surface to its projection area). The equation indicates that the surface roughness can alter the actual wettability of the surface. If $r>1$, then the microstructured surface can increase the hydrophobicity (if $\theta>90^{\circ}$) and hydrophilicity (if $\theta<90^{\circ}$) of the actual smooth surface.

In Cassie's model, due to the presence of air molecules present between the surfaces of two phases (i.e., within roughness introduced channels), solid surface is not perfectly wetted by liquid (**Figure 2c**). Hence, the contact angles of structured (rough) surface (θ_c) and smooth surface (θ) are related by following equation:

$$\cos\theta_c = f_1 \cos\theta - f_2 \tag{17}$$

where f_1 and f_2 are the ratios of solid-liquid interface and solid-gas interface areas to their combined (f_1+f_2) projection area, respectively. According to the equation, the presence of air pockets between the solid and liquid surfaces leads to increase in contact angle value. In addition to absorbent properties, the absorbate's physiochemical characteristics are also crucial in determining the absorption effectiveness of the solid absorbent. Since absorption is a mass transfer phenomenon, it heavily depends upon the density and viscosity of the absorbates. Generally, less viscous liquids (low cohesive energy) tend to absorb faster compare to highly viscous liquids [104]. As the pore volume of a given absorbent is fixed, and hence, it can only absorb a fixed volume of a liquid, the absorption capacity an absorbent can be linearly related with the densities of the absorbing liquids [105].

4. Synthesis of graphene-based hybrids and composites

A number of strategies have been used for the synthesis of variety of adsorbents and absorbents constituted by graphene-based hybrids and composites.

4.1. Dip-coating process

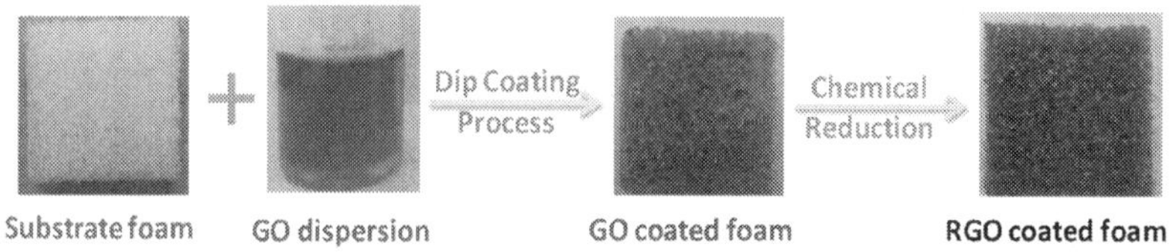

Figure 3. Schematic representation of formation of graphene-coated absorbent sponges by dip-coating process.

The dip coating is actually a two-step process (**Figure 3**) involving coating of graphene oxide (GO) over substrate (foam/particulate-solid) followed by chemical reduction of coating to reduced GO (RGO). In the typical coating process, sponge/foam substrate is immersed inside aqueous dispersion of GO for specified time and dried to form GO-coated foam. The concentration of GO solution and number of dip-dry cycles may be varied to change the coating thickness and uniformity and ultimately the GO loading. Once the desired coating/loading is achieved, the foam may be chemically reduced by exposure to chemical reducing agents such as hydrazine, ascorbic acid or other systems, resulting in the formation of RGO-coated foam. The formation of GO and RGO coatings are evident from the change in the shade of foam to brownish and blackish, respectively. The dip-coating process involved simple physical interactions between GO/RGO coating and foam and possess disadvantages in terms of coating uniformity and its poor adherence. In it modified version involving chemical grafting, suitable functionalized derivative of GO is chemically grafted over substrate (foam/particulate-solid) carrying complimentary functional groups. Such foams display better coating uniformity and adherence compared to dip-coating process.

4.2. Surface coating by *in situ* generation of graphene via thermal treatment

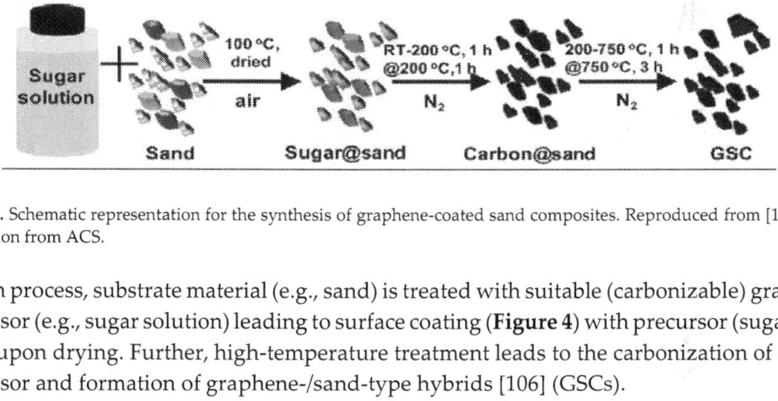

Figure 4. Schematic representation for the synthesis of graphene-coated sand composites. Reproduced from [106] with permission from ACS.

In such process, substrate material (e.g., sand) is treated with suitable (carbonizable) graphene precursor (e.g., sugar solution) leading to surface coating (**Figure 4**) with precursor (sugar with sand) upon drying. Further, high-temperature treatment leads to the carbonization of coated precursor and formation of graphene-/sand-type hybrids [106] (GSCs).

4.3. *In situ* incorporation process

In this process, the aqueous dispersion of GO (colloidal solution) is mixed with suitable secondary phase (CNTs, inorganic nanoparticles such as Fe_{23}/Fe_3O_4) and subjected to further processing to form 3D self-assembled structures and cause reduction of GO phase to RGO.

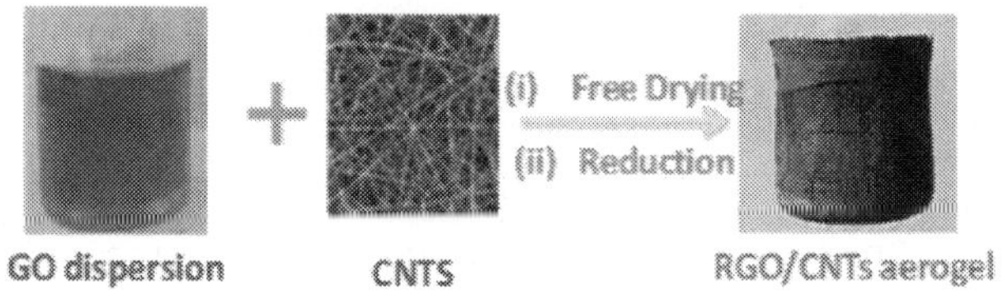

Figure 5. Schematic representation for the synthesis of a graphene composite by *in situ* incorporation of CNTs in GO solution.

For example, GO/CNTs mixed solution is subjected to freeze drying followed by chemical reduction to form RGO/CNTs hybrid-based aerogels (**Figure 5**). The presence of CNTs prevents the stacking of RGO phase and provides improved porosity and mechanical strength to the aerogel.

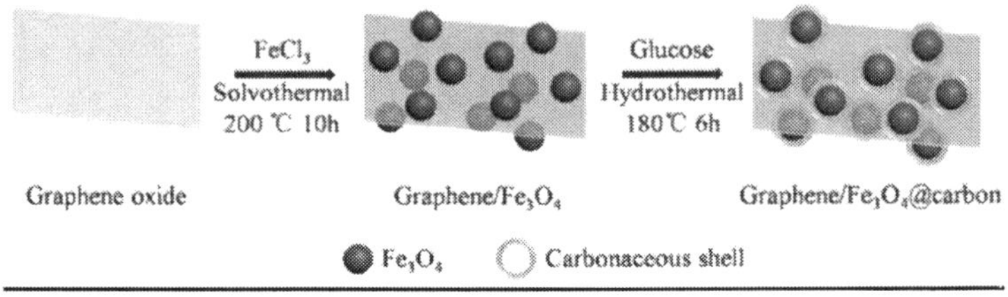

Figure 6. Schematic representation of *in situ* incorporation process involving the mixing of GO with suitable filler followed by secondary processing. Reproduced from [53] with permission from RSC.

Similarly, GO/Fe_3O_4/glucose system is hydrothermally treated at high temperature to form the RGO/Fe_3O_4-based hybrids in a single step (**Figure 6**). Here, the Fe_3O_4 phase checks the RGO agglomeration, improves porosity and allows magnetic separation of the hybrids from the water (after the completion of uptake of pollutant).

4.4. Hydrothermal process in the presence of gelation-causing agent

This involves the hydrothermal treatment of GO dispersion in the presence of suitable gelation agent, for example, chitosan leading to the formation of hydrogel [107], which may lyophilized or supercritical dried to form the corresponding hybrid aerogel.

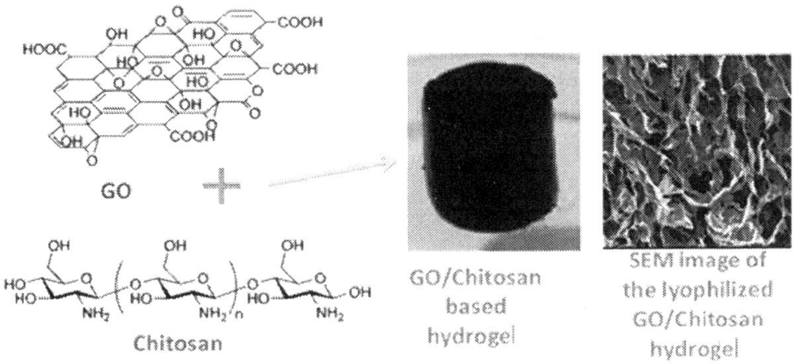

Figure 7. Photograph of GO/chitosan hydrogel and its SEM image. Reproduced from [107] with the permission from RSC.

For example, mixed aqueous dispersion of GO and chitosan may be hydrothermally converted into 3D self-assembled macroscopic hydrogel (**Figure 7**), which consists of highly porous structure as evident from the scanning electron microscope (SEM) investigations of its lyophilized version.

5. The water purification performance of graphene-based hybrids and composites

As already discussed in previous section, due to its high specific surface area, tunable morphological features, controllable porosity, inherent hydrophobicity and oleophilicity, chemical manipulability, tunable surface functionality, facile processing via chemical route and good thermal/environmental/chemical stability, graphene has been widely used to design variety of sorbents to selectively remove the pollutants via adsorption (e.g., for metal ions or organic dyes) or absorption (e.g., for oils or organic solvents) processes [15,17,18,43,53,79,81,91,97,105,108–111]. The contaminant capture ability, including the affinity, toward particular type of contaminant, capturing mechanism and actual purification performance of synthesized graphene-based platforms depends upon the presence of specific set of above properties. This section provides comprehensive account of the purification performance of the graphene-based hybrids and composites for three major classes of pollutants viz. metal ions, organic dyes and oils/organic solvents, with emphasis on the role of above-mentioned specific properties and their correlation to observed performance [47,52,106,112].

5.1. Adsorption of metal ions

It has been well established that the presence of excessive amounts of heavy metals such as Hg (mercury), Pb (lead), As (Arsenic), Cr (chromium), Cu (copper), Ni (nickel), Cd (Cadmium), Se (selenium), etc. in water has a catastrophic effects on living systems and as they degrade the quality of water for other practices like in industries and agricultural sector. The main sources of these metal ions in water systems include effluents of various processes such as galvanization, batteries' waste disposal, mining operations and paper/textile industries, tanneries, etc. [74,91]. Therefore, the removal of above metal ions from the wastewater streams is very important and several efforts in the direction are made using graphene-based adsorbents.

Chandra et al. synthesized magnetite (Fe_3O_4)/reduced graphene oxide (M-RGO) composites [17] and demonstrated their good binding affinity toward As(III) and As(V) ions. The presence of Fe_3O_4 phase enable easy removable of spent adsorbent from water after completion of adsorption of As ions. The composites displayed arsenic removal efficiency (>99.9%) with maximum adsorption capacity (**Figure 8b**) of 13.10 mg/g and 5.83 mg/g for As(V) and As(III), respectively.

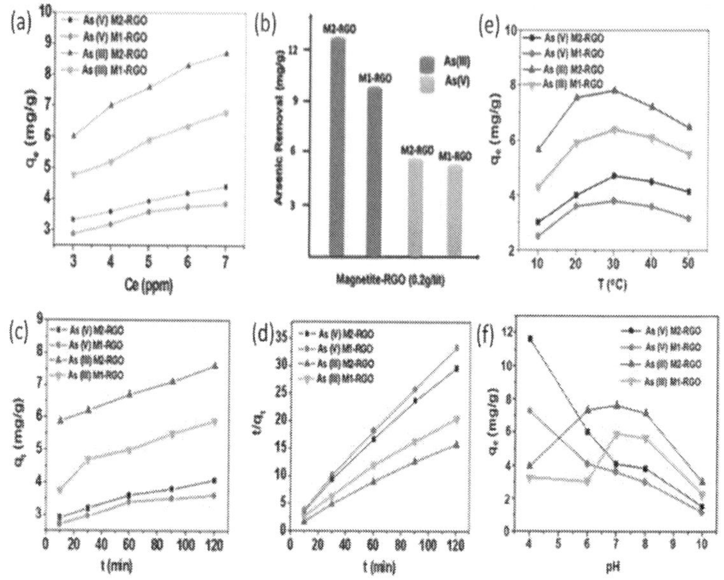

Figure 8. (a) Adsorption isotherms of As(III) and As(V) on the Fe_3O_4-RGO composite (temperature 20°C, pH 7). (b) Maximum adsorption capacity of M-RGO composites for arsenic removal from water. Kinetic absorption data plots of arsenic ions by M-RGO: (c) arsenic removal rate q_t versus time t and (d) the transformed rate plot t/q_t versus t. (e) Effect of temperature on arsenic adsorption: pH, 7; adsorption time, 2 h; adsorbent dose, 0.2 g/L; and arsenic concentration 5 ppm, (f) Effect of pH on arsenic adsorption: temperature, 20°C; adsorption time, 2 h; adsorbent dose, 0.2 g/L; and arsenic concentration, 5 ppm. Adopted from [17] with permission from ACS.

The authors plotted the adsorption isotherm (**Figure 8a**) and fitted kinetic data (**Figure 8c** and **8d**) into pseudosecond-order kinetic model and showed the applicability of the model on adsorption process. They also outlined the optimum temperature (**Figure 8e**) for maximum adsorption capacity. The difference in adsorption mechanism of As(V) and As(III) on M-RGO composites was observed by studying the effect of solution pH (**Figure 8f**) on adsorption capacity. An increase in the adsorption of As(III) (present as H_3AsO_4 arsenous acid) in basic conditions ruled out the electrostatic interactions between positively charged adsorbent and negatively charged arsenious acid. Rather, surface complexation of As(III) ions is responsible for very high adsorption capacity in alkaline conditions. On the other hand, As(V) ions present as $[H_2AsO_3]^-$ species) are electrostatically adsorbed by the adsorbent surfaces at low pH values. Wang et al. [74] exploited the surface area and charge density of graphene oxide sheets in the removal of cadmium and cobalt ions. They have studied the sorbing properties of graphene oxide nanosheets for the removal of Cd(II) and Co(II) ions from aqueous solutions at different pH values and showed that the adsorption of the Co(II) ions increases with increase in the pH of solution reaching maximum at pH>9. In case of Cd(II) ions, 98% adsorption of ions occurred at pH>9 mainly due to increase in negative surface charge density of graphene nanosheets (with an increase in the pH of solution) [74], which electrostatically adsorb the positively charged metal ions. Expanding the work in this area, Fan et al. [113] fused the acid resistant, magnetic cyclodextrin with GO nanosheets (MCGN) for enhancing the adsorption capacity of the composite for Cr(VI) ions (**Figures 9** and **10**).

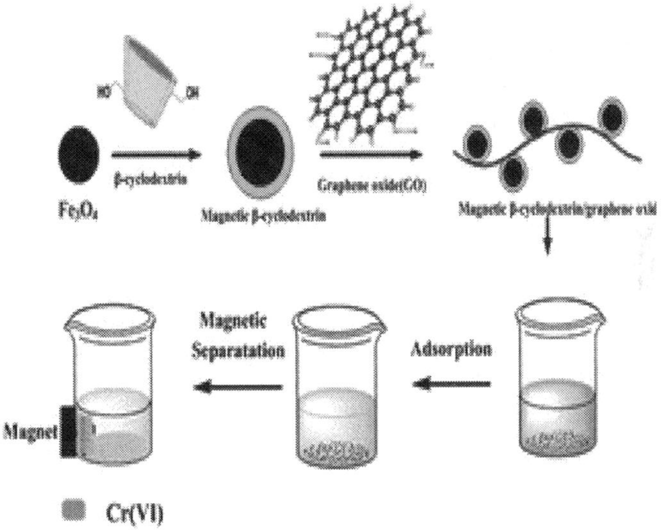

Figure 9. A schematic representation of the preparation of cyclodextrin-graphene composite and removal of Cr(VI). Reproduced from [113] with permission from RSC.

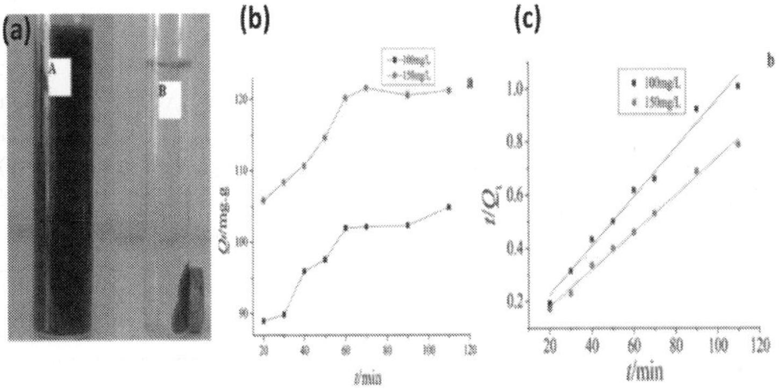

Figure 10. (a) MCGN in the absence (A) and presence (B) of external magnetic field (b) time profile and (c) Kinetics of Cr(VI) ions removal with 1 g/L MCGN concentration and initial Cr(VI) concentrations are 100 mg/L and 150 mg/L. Reproduced from [113] with the permission from RSC.

The adsorbents work better under slightly acidic conditions with good recyclability, making it a good choice for heavy metal ions removal as shown in **Figure 11**.

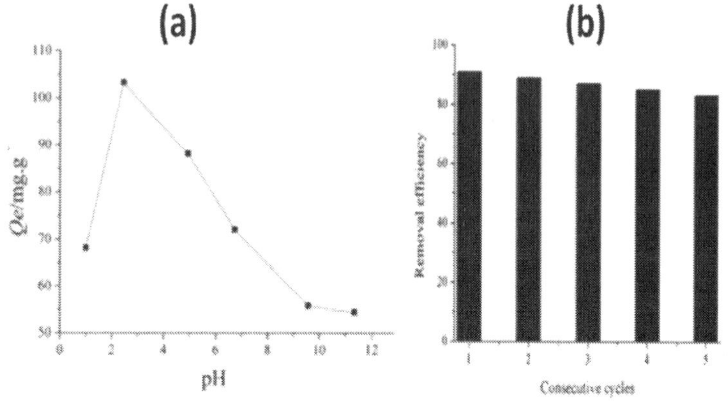

Figure 11. (a) Effect of pH on adsorption capacity of cyclodextrin-graphene nanocomposite. ([CGN] = 1 gL^{-1}, [Cr(VI)] = 100 mgL^{-1}, contact time: 60 min). (b) Recyclability of the cyclodextrin/graphene. Reproduced from [113] with the permission from RSC.

The selectivity of adsorbent for a specific adsorbate is highly desirable. Hao et al. [114] synthesized a SiO_2-graphene composite and specifically adsorb Pb(II) ions from solution. The authors observed that adsorption capacity is highly dependent on the ionic strength of solution and decrease with ionic strength due to completion between various ions for adsorbing sites. Luo et al. [82] highlighted the performance of heavy metal salts such as ZrO_2 hybridized with graphene oxide for the removal of As(III) and As(V) ions. This composite showed ionic strength independent adsorption behavior with high adsorption capacities of 95.15 mg/g and 84.89 mg/g for As(III) and As(v) ions, respectively, with good recyclability. Ren et al. [81] outlined the role of temperature of solution in regulating adsorption. The graphene-MnO_2 composite prepared by them for the removal of Ni(II) ions increase its adsorption capacity by a factor of 0.7 upon changing temperature from 298 K to 318 K. The kinetic parameters obtained from **Table 1** suggested that the pseudosecond-order kinetic model is more appropriate than pseudofirst-order kinetic model. Langmuir adsorption isotherms parameters are given in **Table 2** at various temperatures, which shows the increase in adsorption capacity at higher temperatures.

Adsorbents	(q_e)exp (mg g^{-1})	Pseudofirst-order equation			Pseudosecond-order equation		
		(q_e)cal (mg g^{-1})	K_1 (min^{-1})	R^2	(q_e)cal (mg g^{-1})	K_2 (g $mg^{-1}min^{-1}$)	R^2
GNS/MnO_2	42.855	28.372	0.00072	0.8672	46.447	0.00149	0.9993

Table 1. The kinetics constants of the Ni (II) adsorption onto GNS/MnO_2 composite.

Temperature (K)	GNS/MnO_2		
	q_{max} (mg g^{-1})	K_L (L mg^{-1})	R^2
298	46.55	0.333	0.995
308	60.31	0.310	0.991
318	66.01	0.558	0.999

Table 2. The Langmuir isotherm parameters of the Ni (II) adsorption onto GNS-MnO_2 composite at various temperature (K).

A systematic study was done by Liu et al. [45] for the adsorption of Co(II) ions on magnetite-graphene composite (**Figure 12**).

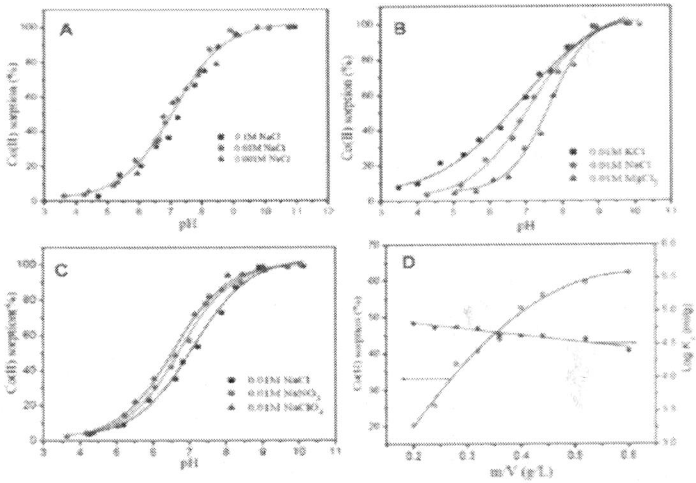

Figure 12. Effect of ionic strength (A), foreign cations (B), foreign anions (C), and sorbent content (D) on Co(II) sorption onto the M-GO composite at pH=6.8, T=303.15 K, I=0.01 M NaCl, concentration of cobalt = 10.0 mg/L from [45] with the permission from ACS.

The effect of ionic strength on adsorption capacity is negligible as shown in **Figure 12a**. But the presence of foreign ions significantly affected the adsorption behavior of the adsorbent. The pH of solution higher than 8.5 was favorable of maximum adsorption capacity. Zhou et al. [79] also reported the dependence of adsorption capacity of graphene composites on pH of solution. The material showed decreased affinity for Cr(IV) ions at higher pH as hydroxyl ion compete with the CrO_4^{2-} ions for adsorbing sites [5]. Such studies are also reported by other groups to show how metal ion compete for adsorbing sites in the environment of excessive hydroxyl ions and hydronium ions [115]. It can be concluded from the discussion earlier that graphene-based materials are highly promising for the removal of heavy metal ions from wastewater systems.

5.2 Adsorption of dyes

Synthetic organic dyes that represent important color imparting ingredient of leather, paper, textile and rubber industries are also present in significant amount in their wastewater streams [22]. These dye stuffs have serious health hazards for both humans and aquatic systems [21,116]. In recent times, graphene analogs, including GO- or RGO-based sponges, composites and hybrids, have been successfully employed for the removal of toxic dyes from wastewater.

Liu et al. [47] used a centrifugal evaporator (**Figure 13a**) to synthesize GO-based sponge (**Figure 13b**) consisting of 3D interconnected network of graphene layers and demonstrated their potential for the removal of cationic dyes such as methylene blue (MB) and methyl violet (MV) with removal efficiency of 99.1 and 98.8%, respectively, with equilibrium reaching within 2 min. It has been highlighted that the microporous structure of GO sponge (**Figure 13c**), high negative charge density (due to rich oxygen functionalities) mediated cation-anion-type interactions, π-π stacking interactions between dye backbone and adsorbent and high specific surface area (148.4 m^2/g), collectively contribute toward the excellent dye adsorption capacity (**Figure 13d**), that is, 397 mg/g and 467 mg/g for MB and MV, respectively (**Figure 13**).

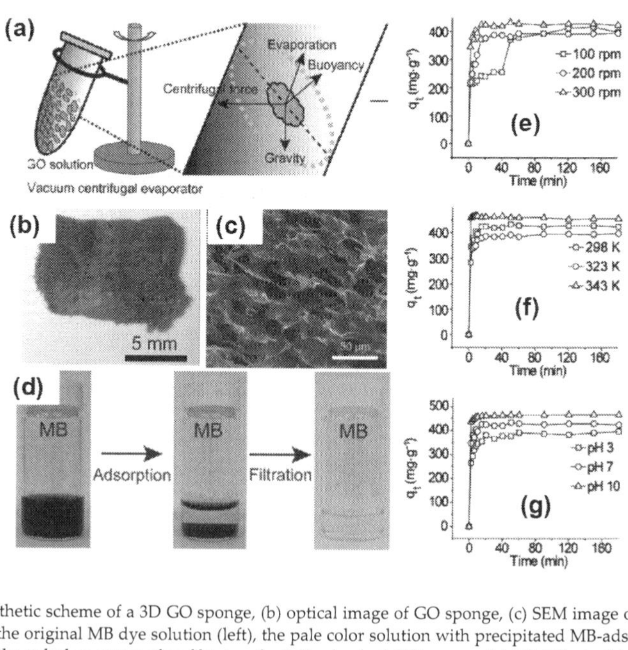

Figure 13. (a) Synthetic scheme of a 3D GO sponge, (b) optical image of GO sponge, (c) SEM image of GO sponge, (d) digital images of the original MB dye solution (left), the pale color solution with precipitated MB-adsorbed GO sponges (middle), and the colorless water after filtering the MB adsorbed GO sponge (right). Effect of (e) stirring rate, (f) temperature and (g) pH on the adsorption rate and capacity for MV dye. Reprinted from [47] with permission from ACS.

The effect of stirring (**Figure 13e**), temperature (**Figure 13f**) and pH (**Figure 13g**) on the adsorption rate and capacity have also been investigated, with the observation of increase in these parameters favoring adsorption rate and capacity. Once the adsorption is complete, spent adsorbent has to be removed from purified water by filtration process, which is not only bit cumbersome but also involve risks of recontamination. Gupta et al. [106] prepared sugar-derived graphene sand composite (GSC) consisting of graphene-coated sand particles (**Figure 14a** and **b**) via thermal processing. This cost-effective active adsorbent material-packed column (**Figure 14c**) showed good R6G dye removal efficiency with adsorption capacity of 50–55 mg/g. The time-dependent adsorption studies were also carried out to understand the system kinetics (**Figure 14**).

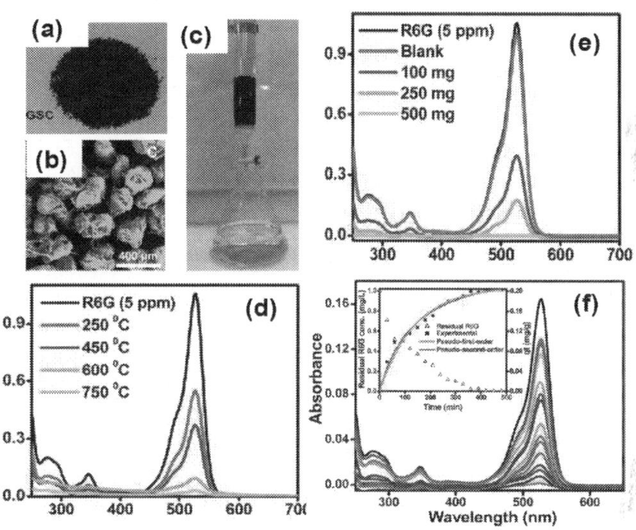

Figure 14. (a) Digital image and (b) SEM image of GSC, (c) photograph of GSC750 packed adsorption column for the separation of R6G dye. Variation of adsorption capacity as a function of (d) GSC heating temperature and (e) GSC content. (f) Kinetic study of dye adsorption with 30-min time interval. Inset shows the pseudofirst- and pseudosecond-order model fits. Reproduced from [106] with permission from ACS.

The authors highlighted the dependence of adsorption characteristics on GSC-heating temperature (**Figure 14d**) and adsorbent content (**Figure 14e**). The kinetic data were examined in terms of Lagergren pseudofirst-order (Section 2.5.1) and Ho's pseudosecond-order (Section 2.5.2) kinetic models [95] and the fitted data (inset of **Figure 14f**) showed that a pseudosecond-order equation is more appropriate in describing the experimental data. Li et al. [84] synthesized a $Mg(OH)_2$-RGO composite(GMC) to improve the adsorption capacity by increasing the porosity of the adsorbent. They pointed that surface charge density and π conjugation play an important role for high removal efficiency of GMC for cationic dyes such as MB. The presence of $Mg(OH)_2$ ensured the formation of mesoporous structure (**Figure 15** calculation of BET surface area and pore size distribution) and inhibit restacking of graphene sheets, thereby generating high accessible surface area (|104 m^2/g) for adsorption (**Figure 15**).

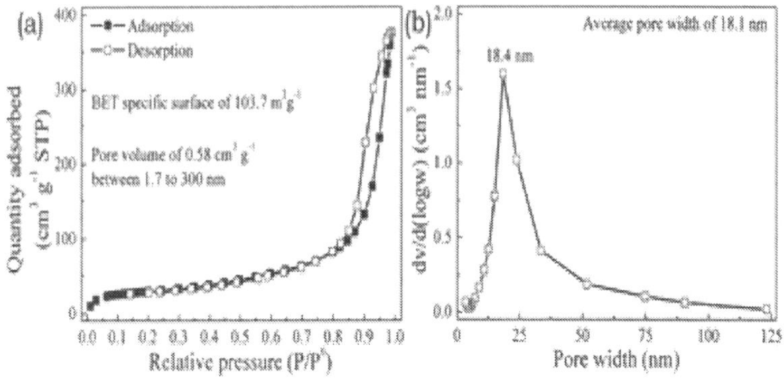

Figure 15. (a) N_2 adsorption-desorption isotherm at 77 K and (b) pore width distribution of MGC. Adopted from [84] with permission from RSC.

These GMC readily desorbs the MB molecules in ethanol solution reflecting promising recyclability. It has been observed that despite their good adsorption capacity, the separation of these graphene-based powdered adsorbents is a difficult and cumbersome task. Therefore, they pose inherent disadvantages in terms of requirement of energy-intensive separation and regeneration techniques along with chances of recontamination of water with any leftover adsorbent particles. In order to overcome poor separation ability of powdered absorbents, Fan et al. [53] synthesized graphene-Fe_3O_4 with carbon (GFC)-type hybrid materials with coupled porosity and magnetic properties for MB removal (**Figure 16**).

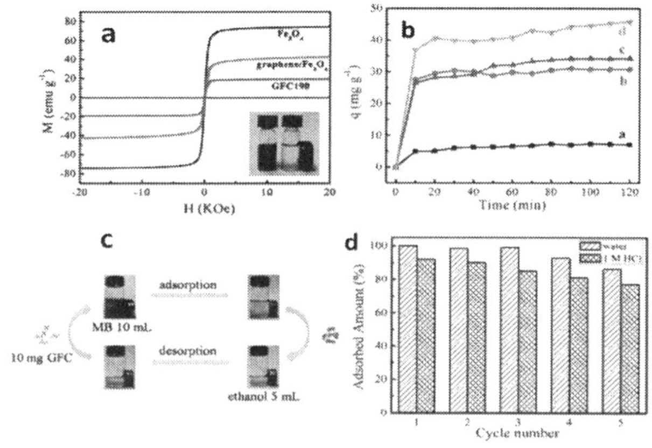

Figure 16. (a) Magnetic hysteresis loops for pure Fe_3O_4, graphene-Fe_3O_4 and GFC hybrids, (b) adsorption curves for pure Fe_3O_4, graphene-Fe_3O_4, Fe_3O_4@carbon hybrid, GFC hybrid, (c) schematic representation of recyclability of GFC hybrid, and (d) percentage removal of MB by GFC hybrid in water and 1 M HCl solution in five cycles. Reproduced from [53] with permission from RSC.

This material exhibits good porosity and magnetic character (**Figure 16a**) resulting in good dye adsorption response (**Figure 16b**) with good capacity (73.26 mg/g) and fast adsorption rates (i.e. removal of 80% dye within 20 min). The superiority of such materials lies in the fact that they can be easily separated after adsorption of dye from water using an external magnetic field and can be regenerated by desorbing the dye in ethanol (**Figure 16c**). The good recyclability is evident from very low decrease in adsorption capacity after five cycles of regeneration and reuse. The adsorption mechanism of MB on GFCs was examined by applying Langmuir and Freundlich isotherm models and their parameters are reported in **Table 3**.

Langmuir				**Freundlich**		
Adsorbent	q_{max} **(mg g^{-1})**	b **(L mg^{-1})**	R^2	K	N	R^2
GFC190	73.26	0.58	0.96	31.74	3.66	0.92

Table 3. Langmuir and Freundlich parameters for the adsorption of MB on GFC hybrid.

On the basis of regression coefficient (R^2) values, Freundlich adsorption is ruled out and result better fits into the Langmuir adsorption model. The good value of obtained maximum adsorption capacity (i.e., 73.26 mg/g) has been ascribed to high surface area of GFCs, which enhances the accessibility of diffusion pathways for dye molecules in macroporous domains. Additionally, strong electrostatic interactions and π-π interactions between the aromatic cationic dye and negatively charged GFC hybrid are mainly responsible for rapid adsorption rates. Wang et al. [108] also synthesized a Fe_3O_4 -graphene-based magnetic composite for the removal of fuchsine dye and showed an excellent adsorption capacity of 89.4 mg/g in very shorter time period. The kinetics of adsorption was determined by fitting the results to (**Figure 17a**) pseudosecond-order kinetic model (Eq. (13)). The regression coefficient value of 0.9999 shows the applicability of the model and confirmed the chemisorption of dye molecules on adsorbent surface (**Figure 17**).

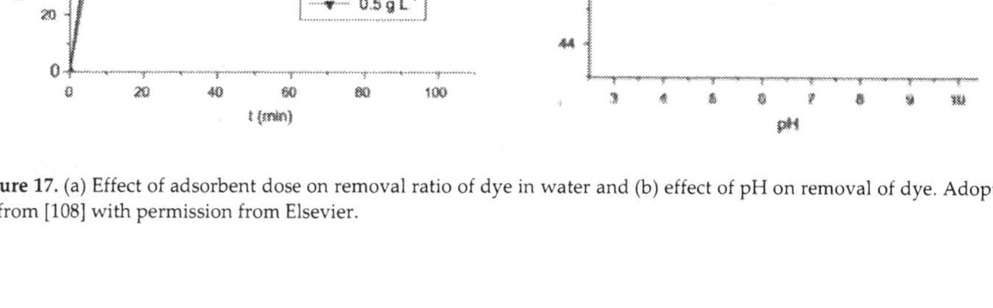

Figure 17. (a) Effect of adsorbent dose on removal ratio of dye in water and (b) effect of pH on removal of dye. Adopted from [108] with permission from Elsevier.

The authors also studied the role of dye solution pH (**Figure 17b**) on the adsorption mechanism of the composite. The adsorption capacity was found to increase with pH (from 3 to 5.5) mainly due to the fact that at low pH condition protonation of basic dye (amido groups) fuchsine occurs, which leads to increase solubility of dye in water and hence its relatively poor adsorption under slightly acidic conditions. The extensive research on development of such handy magnetic graphene materials is now of great interest as these materials have high dye removal efficiency, quick dye adsorption and facile removal of spent adsorbent from purified water after completion of pollutant adsorption. In this series, a very promising magnetic adsorbent Ni-RGO (NGC) composite is prepared by Li group [117], which almost completely removes rhodamine (RhB) and MB dyes from water within 4 h as evident from corresponding time-dependent UV-visible spectra (**Figure 18**) (**Table 4**).

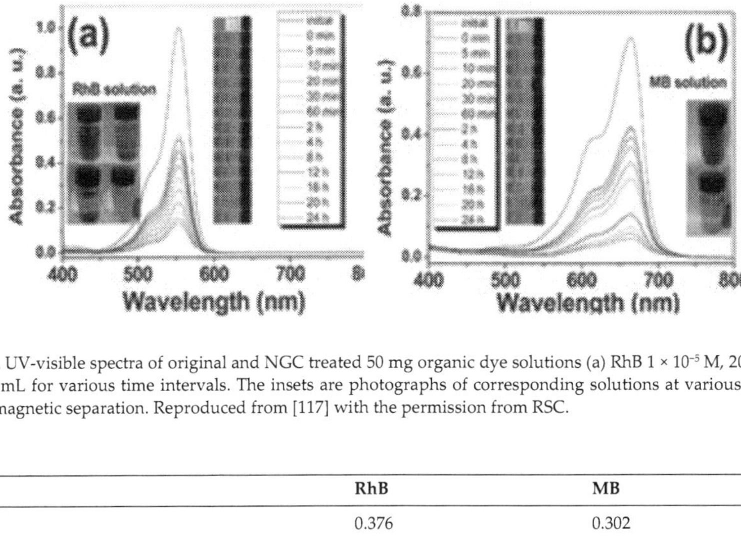

Figure 18. UV-visible spectra of original and NGC treated 50 mg organic dye solutions (a) RhB 1 × 10^{-5} M, 20 mL (b) 2 × 10^{-5}M, 20 mL for various time intervals. The insets are photographs of corresponding solutions at various time intervals and magnetic separation. Reproduced from [117] with the permission from RSC.

Constant	RhB	MB
K	0.376	0.302
N	1.048	0.894
R^2	0.99	0.99

Table 4. Freundlich isotherm constants and their correlation coefficients.

The authors show that the Freundlich adsorption mode is applicable to NGC with regression coefficient value (R^2) of 0.99 for both RhB and MB dyes, which indicates the heterogeneity of adsorbent surface. The Freundlich exponent (n) for RhB (1.048) is greater compared to MB (0.894) that shows the better interaction of the RhB dye with adsorbent compared to MB. This is mainly due to molecular structural differences between the two dyes. In particular, MB has a soft S-based cationic system and RhB has a hard N-based cationic system, resulting in high affinity of latter for a hard O-based anionic center. Jiang et al. [110] synthesized a magnetic chitosan-graphene oxide (MCGO) nanocomposite for the adsorption of anionic dye methyl orange (MO). The magnetic adsorbent showed extraordinary MO adsorption capacity (398.08 mg/g) and easily separated by magnetic assistance after completion of dye adsorption **Figure 19a**.

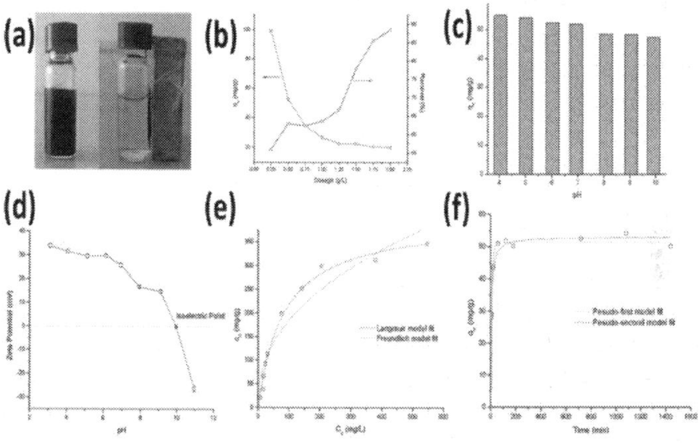

Figure 19. (a) Digital photograph of MO adsorption on MCGO nanocomposite and its magnetic separation. Variation in adsorption capacity with (b) adsorbent mass and (c) solution pH. Effect of (d) pH on zeta potential of the MCGO nanocomposite, (e) Experimental data and curve fits for Langmuir and Freundlich isotherms (f) contact time on MO adsorption by MCGO nanocomposite. Reproduced from [110] with permission from International Journal of Biological Macromolecules.

The authors showed that in case of MO, adsorption capacity decreases with pH of solution as the negatively charged MO molecules compete with hydroxide ions in alkaline conditions shown in **Figure 19c**. Also, zeta potential measurement **Figure 19d** showed that the MCGO nanocomposite attained isoelectric point at pH 10 and hence has poor affinity for anionic dye under high alkaline conditions. Because of high concentration gradient between adsorbate and adsorbent species, rapid dye adsorption takes place initially followed by a gradual decrease in adsorption rate (**Figure 19f**) as at this stage most of the adsorption sites are occupied by the dye molecules. The model fits of Langmuir and Freundlich isotherms for MO are shown in **Figure 19e**. It was found that the Langmuir isotherm better described the adsorption of MO with the higher regression coefficient R^2 (0.9897) than Freundlich isotherm (R^2=0.9112). The kinetic study concluded the applicability of the pseudosecond-order kinetic model for dye adsorption over pseudofirst-order kinetic equation and curve fitting of the experimental data for the two models is shown in **Figure 19f**.

5.3. Absorption of oils/organic solvents

Addressing the severe water pollution arising from oil spillage and chemical leakage is a challenging task. In principle, materials with both superhydrophobic and superoleophilic properties are highly desired for solving the environmental issues because of their capacity of selective absorption of oils and organic solvents while repelling water completely. As a remediation of accidents, the cleanup of leaked chemicals or oil is a global environment issue. Generally, porous materials that have high porosity and high surface areas show great potential in oil/chemicals pickup; however, these materials usually show poor selectivity toward target oil, thus easily saturated with water. Recently, graphene-based 3D porous structures, possessing superhydrophobic and superoleophilic surfaces, attracted enormous attention owing to their high affinity for various oils and organic solvents and ability to selectively absorb them from their dispersion, emulsions or floating layered system with water. In particular, some work has been done in the direction of graphene-coated polymeric foams [105] type hybrids and graphene/composites based aerogels [118]. A number of polymeric foam substrates have been exploited for coating/decoration with graphene, including polyurethane (PU), polydimethoxysiloxane (PDMS) and melamine foams.

Liu et al. [105] fabricated reduced graphene oxide-coated polyurethane sponges (rGPU) via a two-step process involving dip coating with GO followed by its chemical reduction to RGO. In this system, porous structure of PU sponge-based substrate (SEM image **Figure 20a**) is coated with RGO layers (SEM image **Figure 20b**) without disturbance of macroporous cellular architecture of the sponge. However, due to its inherent hydrophobic nature, the uniformly coated of RGO impart high degree of hydrophobicity and oleophilicity to PU sponges, as confirmed from the contact angle measurements for water (**Figure 20c**) and oil (**Figure 20d**).

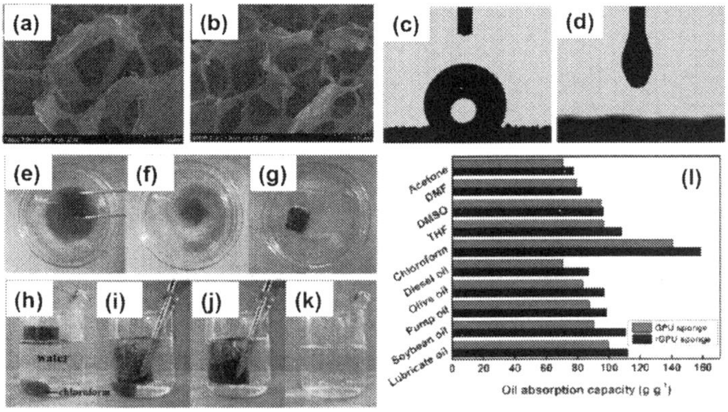

Figure 20. (a, b) SEM images of RGPU and (c, d) optical image of a water drop and lubricating oil drop on RGPU surface. Digital images of oil (e–g) and chloroform (h–k) absorption. Absorption capacity of RGPU sponge for various oils. Reproduced from [105] with the permission from ACS.

Such a combination of macroporosity, hydrophobicity and oleophilicity of these materials leads to efficient absorption of lubricating oil (**Figure 20e–g**), chloroform (**Figure 20h–k**) and variety of other oils and organic solvents with good specific absorption capacity.

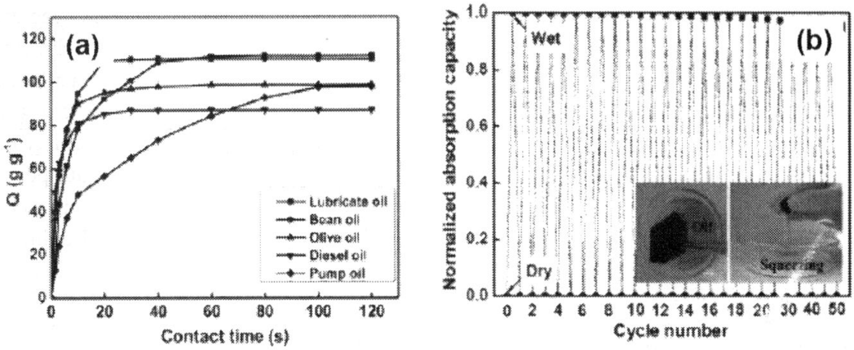

Figure 21. (a) Absorption capacities of RGPU sponge with contact time with various oils and (b) recyclability of RGPU sponge. Reproduced from [105] with the permission from ACS.

In addition, these hybrid sponges also display high absorption rate (**Figure 21a**) with saturation attaining in 2 h, The authors also exploited the compressible nature of PU sponge for the reusability of these materials and shown that they can be reused number of times (**Figure 21b**) without much loss in absorption capacity.

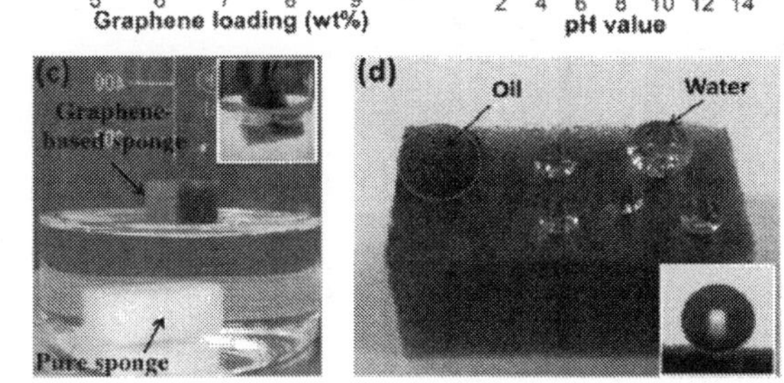

Figure 22. Effect of graphene loading on the water CA of the pure sponge (a). Relationship between pH values and CAs of the graphene-based sponges after 12 h floating on aqueous solutions with different pH values (b). Photograph of pure (white color) and graphene-based (black color) sponges after being placed on water, the inset is a photograph of the graphene-based sponge partially immersed in water by a force (c). Water droplets as quasi-spheres and motor oil trace on the surface of the graphene-based sponge (d), the inset in (d) is an optical image of a water droplet at a CA of 162. Reproduced from [86] with permission from RSC.

Nguyen group prepared graphene-coated melamine sponges [86] and highlighted the role of graphene loading and solution pH on the water contact angle value of synthesized hybrid foams. The presence of graphene nanolayers on the surface of melamine sponge drastically changed its surface properties attributed by the fact that the water contact angle changed from 0° (for pristine melamine sponge) to 162° (for graphene-coated melamine sponge). Further, the stability of superhydrophibicity to aggressive environment is demonstrated by the fact that the WCA remained greater than 150 degree even after 12 h of floating over aqueous solutions of different pH values. The pure sink to the bottom (**Figure 22c**), whereas the graphene-coated sponge floats over water, retard forced immersion with no water uptake, which further demonstrates its superhydrophobic nature. The silver mirror-like appearance of the immersed part of the graphene-based sponge (the inset **Figure 22c**) is attributed to a uniform air layer trapped between the water and superhydrophobic surfaces, which are referred to nonwetting Cassie-Baxter surfaces (Section 3). Interestingly, the sponge also displays superoleophilicity such that water droplets (pointed arrows) display WCA of 162° (inset **Figure 22d**) and remained unabsorbed (**Figure 22d**), whereas motor oil droplet (encircled) was immediately absorbed. The coexistence of superhydrophobic and superoleophilic surfaces is attributed to the combination of the microporous structure (with trapped air pockets) of the melamine sponge substrate and the hydrophobicity of the graphene nanosheets, and the micro-/nanotextured structure of graphene nanosheets on sponge skeletons.

These hybrid sponges display good absorption capacities (ranging from 60 to 165 times of its own weight) for motor oil (**Figure 23a**), chloroform (**Figure 23b**) and several other solvents (**Figure 23c**), good recyclability, low density and chemical inertness.

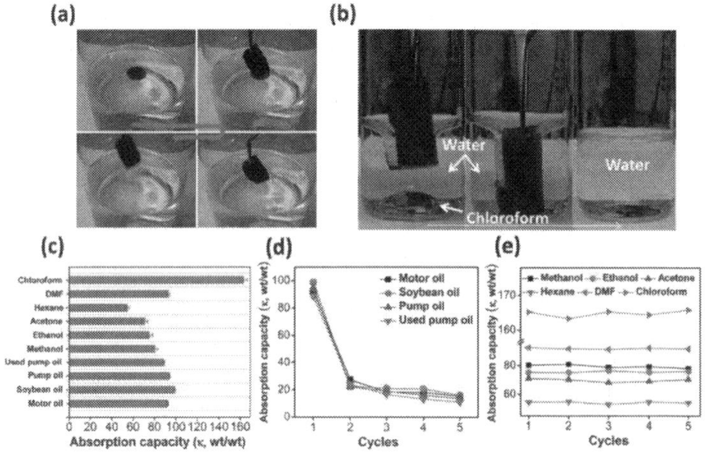

Figure 23. Snapshots of the removal process of motor oil (dyed with solvent red 24 for clear observation) (a) and chloroform (b) from water using the graphene-based sponges. Oil and organic solvent absorption capacities of the graphene-based sponges (c). Absorption recyclability of the graphene-based sponges for oils (d) and organic solvents (e). Reprinted from [86] with permission from RSC.

The recyclability and reusability of the graphene-based sponge also plays an important role in oil/chemical clean up applications. In general, sponge regeneration and absorbed pollutant separation can be realized either by squeezing the sponges for high-boiling materials like oils or by distillation/squeezing for low-boiling oils/solvents. It is observed that the generalized recyclable absorption response is similar for all tested oils (**Figure 23d**) with large decrease in absorption capacity after first cycle due to the residual oils inside the sponges, which cannot be removed by manual squeezing. After the second cycle, the saturated absorption capacity decreases only slightly indicating good recyclability. However, in case of organic solvents, desorbed sponges (after distillation) display no obvious change in saturated absorption capacity (**Figure 23e**) reflecting complete pollutant (solvent) recovery and good sponge recyclability, which are requisite for practical utility. Despite their great promise for oil/chemical clean up, the poor adhesion between coated graphene and foam substrate, remained a problematic issue. An attempt has been made by Tajendra et al. to improve the adhesive forces between PU sponges and RGO via covalent bonding strategy [119] involving use of silane- or titanate-based coupling agents. The sponges exhibited absorption capacity of 30–50 g/g depending upon the nature of oil/solvent. Further, the authors reported that the covalent bonding between RGO and PU prevents the leaching problem of surface layers and hence increase the recyclability of the hybrid sponges.

Table 5 compares the oil/organic-solvent removal performance of various graphene-coated foams based hybrids. It can be seen that the achieved absorption capacities are still lower than theoretically possible values and there is large deviation between results of different works. Further, besides hydrophobicity of the hybrid sponges, the parameters like viscosity, densities, polarities, surface tensions/energies of spilled chemicals and pore size, porosity and surface texturing of the foamed substrates, plays an important role in governing absorption characteristics (absorption capacity, absorption rate, selectivity) and need to be looked into thoroughly to further understanding and improvements. The graphene coating's adherence issues are sorted out to improve the reusability and to reduce the chances of cross contamination. In order to drastically improve the gravimetric absorption capacities, eliminate the adherence issues and improve the recyclability, all graphene-based 3D self-assembled networks such as graphene aerogel seems to be promising solution [16,75].

S. No.	Absorbate	Absorption capacities (g/g) of various materials			
		Graphene-coated melamine sponge	GO-coated PU sponges	RGO-coated PU sponges	TTEAI-RGO sponge
1	Chloroform	165	140	160	35
2	DMF	100	78	82	–
3	Hexane	60	–	–	–
4	Acetone	75	75	80	32
5	Ethanol	78	–	–	–

S. No.	Absorbate	Absorption capacities (g/g) of various materials			
		Graphene-coated melamine sponge	GO-coated PU sponges	RGO-coated PU sponges	TTEAI-RGO sponge
6	Used pump oil	90	–	–	–
7	Methanol	80	–	–	–
8	Pump oil	95	90	100	28
9	Soya bean oil	100	92	110	–
10	Motor oil	90	–	–	–
11	Diesel oil	–	75	90	–
12	Olive oil	–	85	100	–
13	Lubricate oil	–	100	112	–
14	Gasoline oil				45
Ref.		[103]	[63]	[63]	[104]

PU—polyurethane sponges, TTEAI— titanium (IV)(triethanolaminato) isopropoxide.

Table 5. Absorption capacities (g/g) of graphene-based materials for various oils/organic solvents.

Nevertheless, huge efforts in the direction are needed for the realization of efficient, stable, cost-effective and commercially viable absorbents for large-scale oil/chemical clean up to be used at actual spill site especially under marine environment.

7. Conclusion

In summary, graphene-based composites/hybrids have a tremendous potential for water purification applications. The inherent properties of graphene in conjunction with other materials induce additional functionalities in the composites/hybrids reflected in terms of enhanced, surface areas, porosity, flexibility, induced magnetism, improved adsorption/ absorption rates and capacities. As discussed in this chapter, these materials show extraordinary chemical stabilities with good recyclability and durability, which are some of the most crucial attributes of sorbing materials. Importantly, the generation of magnetism in graphene composites/hybrids with magnetic nanomaterials prevents to ease out the sorbent operationality and its recovery process. But issues, such as desorption of pollutants for the regeneration of the active materials, are still a major challenge and need to be thoroughly investigated.

Author details

Rahul Sharma and Parveen Saini*

*Address all correspondence to: pksaini@nplindia.org; parveensaini580@gmail.com

Conjugated Polymer and Graphene Technology Lab, Polymeric and Soft Materials Section, Materials Physics and Engineering Division, CSIR-National Physical Laboratory, New Delhi, India

References

[1] Peter Osei Boama, Jacqueline Onumah, Mohammed Takase, Prince Osei Bonsu & Tayari Salifu. AIR POLLUTION CONTROL TECHNIQUES. *G.J.B.B* 1, 124–131 (2012).

[2] Yang, Y., Gupta, M. C., Dudley, K. L. & Lawrence, R. W. Novel Carbon Nanotube-Polystyrene Foam Composites for Electromagnetic Interference Shielding. *Nano Lett.* 5, 2131–2134 (2005).

[3] Romero, J. V. *et al*. A combinatorial approach to screening carbon based materials for respiratory protection. *J. Hazard. Mater.* 183, 677–687 (2010).

[4] Marchiol, L., Assolari, S., Sacco, P. & Zerbi, G. Phytoextraction of heavy metals by canola (Brassica napus) and radish (Raphanus sativus) grown on multicontaminated soil. *Environ. Pollut.* 132, 21–27 (2004).

[5] Kirkham, M. B. Cadmium in plants on polluted soils: Effects of soil factors, hyperaccumulation, and amendments. *Geoderma* 137, 19–32 (2006).

[6] Khan, F. I. & Kr. Ghoshal, A. Removal of Volatile Organic Compounds from polluted air. *J. Loss Prev. Process Ind.* 13, 527–545 (2000).

[7] Zhang, H., Lv, X., Li, Y., Wang, Y. & Li, J. P25-Graphene Composite as a High Performance Photocatalyst. *Acs Nano* 4, 380–386 (2010).

[8] Zhao, L. *et al*. TiO2–graphene sponge for the removal of tetracycline. *J. Nanoparticle Res.* 17, (2015),17:16

[9] Yang, Y., Yi, H. & Wang, C. Oil Absorbents Based on Melamine/Lignin by a Dip Adsorbing Method. *Acs Sustain. Chem. Eng.* 3, 3012–3018 (2015).

[10] Upadhyay, R. K., Soin, N. & Roy, S. S. Role of graphene/metal oxide composites as photocatalysts, adsorbents and disinfectants in water treatment: a review. *Rsc Adv* 4, 3823–3851 (2014).

[11] Saini,P.(2015) *Fundamentals of Conjugated Polymer Blends, Copolymers and Composites: Synthesis, Properties and Applications*. (John Wiley & Sons, Inc., 2015). Hoboken, NJ, USA. doi: 10.1002/9781119137160.ch1 <http://doi.wiley.com/10.1002/9781119137160>

[12] Saini, P. & Choudhary, V. Enhanced electromagnetic interference shielding effectiveness of polyaniline functionalized carbon nanotubes filled polystyrene composites. *J. Nanoparticle Res*. 15, (2013), 15: 1415

[13] Saini, P. (2013) Electrical Properties and Electromagnetic Interference Shielding Response of Electrically Conducting Thermosetting Nanocomposites, in Thermoset Nanocomposites (ed V. Mittal), Wiley-VCH Verlag GmbH & Co. KGaA, Weinheim, Germany. doi: 10.1002/9783527659647.ch10

[14] Fu, F. & Wang, Q. Removal of heavy metal ions from wastewaters: A review. *J. Environ. Manage*. 92, 407–418 (2011).

[15] Yilmaz, M., Tay, T., Kivanc, M. & Turk, H. Removal of corper(II) Ions from aqueous solution by a lactic acid bacterium. *Braz. J. Chem. Eng*. 27, 309–314 (2010).

[16] Xue, Z., Cao, Y., Liu, N., Feng, L. & Jiang, L. Special wettable materials for oil/water separation. *J Mater Chem* 2, 2445–2460 (2014).

[17] Chandra, V. *et al*. Water-Dispersible Magnetite-Reduced Graphene Oxide Composites for Arsenic Removal. *Acs Nano* 4, 3979–3986 (2010).

[18] Tan, I. A. W., Ahmad, A. L. & Hameed, B. H. Adsorption of basic dye on high-surface-area activated carbon prepared from coconut husk: Equilibrium, kinetic and thermodynamic studies. *J. Hazard. Mater*. 154, 337–346 (2008).

[19] Whitehead, P. G. & Lack, T. Dispersion and self-purification of pollutants in surface water systems. *United Nations Educ. Sci. Cult. Organ. 7 Place Fontenoy 75700 Paris*

[20] El Qada, E. N., Allen, S. J. & Walker, G. M. Adsorption of basic dyes from aqueous solution onto activated carbons. *Chem. Eng. J*. 135, 174–184 (2008).

[21] TOXNET (Toxicology Database Network. at <http://toxnet.nlm.nih.gov/cgi-bin/sis/search/a?dbs+hsdb:@term+@DOCNO+1405>

[22] Sulak, M. T. & Yatmaz, H. C. Removal of textile dyes from aqueous solutions with eco-friendly biosorbent. *Desalination Water Treat*. 37, 169–177 (2012).

[23] Kayode, J., Olowoyo, O. & Oyedeji, A. The Effects of Used Engine Oil Pollution on the Growth and Early Seedling Performance of Vigna uniguiculata and Zea mays. *Res. J. Soil Biol*. 1, 15–19 (2009).

[24] at <https://en.wikipedia.org/wiki/Deepwater_Horizon_oil_spill>

[25] Ordinioha, B. & Brisibe, S. The human health implications of crude oil spills in the Niger delta, Nigeria: An interpretation of published studies. *Niger. Med. J*. 54, 10 (2013).

[26] Dalton, T. & Jin, D. Extent and frequency of vessel oil spills in US marine protected areas. *Mar. Pollut. Bull.* 60, 1939–1945 (2010).

[27] Vazquez-Duhalt, R. Environmental impact of used motor oil. *Sci. Total Environ.* 79, 1–23 (1989).

[28] Bayat, A., Aghamiri, S. F., Moheb, A. & Vakili-Nezhaad, G. R. Oil Spill Cleanup from Sea Water by Sorbent Materials. *Chem. Eng. Technol.* 28, 1525–1528 (2005).

[29] Nwilo, P. C. & Badejo, O. T. OIL SPILL PROBLEMS AND MANAGEMENT IN THE NIGER DELTA. *Int. Oil Spill Conf. Proc.* 2005, 567–570 (2005).

[30] OKOYE & OKUNROBO. IMPACT OF OIL SPILL ON LAND AND WATER AND ITS HEALTH IMPLICATIONS IN ODUGBORO COMMUNITY, SAGAMU, OGUN STATE, NIGERIA. *World J. Environ. Sci. Eng.* 1, 1–21 (2014).

[31] Protecting Our Ocean and Coastal Economies. https://www.nrdc.org/sites/default/files/offshore.pdf dated 24 December, 2015

[32] Joshi, R. K. *et al.* Precise and Ultrafast Molecular Sieving Through Graphene Oxide Membranes. *Science* 343, 752–754 (2014).

[33] Dushenkov, V., Kumar, P. B. A. N., Motto, H. & Raskin, I. Rhizofiltration: The Use of Plants to Remove Heavy Metals from Aqueous Streams. *Environ. Sci. Technol.* 29, 1239–1245 (1995).

[34] Zhang, Y., Tang, Z.-R., Fu, X. & Xu, Y.-J. TiO_2-Graphene Nanocomposites for Gas-Phase Photocatalytic Degradation of Volatile Aromatic Pollutant: Is TiO_2-Graphene Truly Different from Other TiO_2-Carbon Composite Materials? *Acs Nano* 4, 7303–7314 (2010).

[35] Lachheb, H. *et al.* Photocatalytic degradation of various types of dyes (Alizarin S, Crocein Orange G, Methyl Red, Congo Red, Methylene Blue) in water by UV-irradiated titania. *Appl. Catal. B Environ.* 39, 75–90 (2002).

[36] Zhang, X., Liu, D., Yang, L., Zhou, L. & You, T. Self-assembled three-dimensional graphene-based materials for dye adsorption and catalysis. *J Mater Chem* 3, 10031–10037 (2015).

[37] Bi, H. *et al.* Spongy Graphene as a Highly Efficient and Recyclable Sorbent for Oils and Organic Solvents. *Adv. Funct. Mater.* 22, 4421–4425 (2012).

[38] Song, W.-L. *et al.* Tuning three-dimensional textures with graphene aerogels for ultra-light flexible graphene/texture composites of effective electromagnetic shielding. *Carbon* 93, 151–160 (2015).

[39] Wang, C.-F., Huang, H.-C. & Chen, L.-T. Protonated Melamine Sponge for Effective Oil/Water Separation. *Sci. Reports* 5, 14294 (2015).

[40] Broje, V. & Keller, A. A. Improved Mechanical Oil Spill Recovery Using an Optimized Geometry for the Skimmer Surface. *Environ. Sci. Technol.* 40, 7914–7918 (2006).

[41] Goodbody-Gringley, G. *et al.* Toxicity of Deepwater Horizon Source Oil and the Chemical Dispersant, Corexit® 9500, to Coral Larvae. *Plos One* 8, e45574 (2013).

[42] Brooijmans, R. J. W., Pastink, M. I. & Siezen, R. J. Hydrocarbon-degrading bacteria: the oil-spill clean-up crew. *Microb. Biotechnol.* 2, 587–594 (2009).

[43] Zhao, J., Ren, W. & Cheng, H.-M. Graphene sponge for efficient and repeatable adsorption and desorption of water contaminations. *J. Mater. Chem.* 22, 20197 (2012).

[44] Mukherjee, R., Sharma, R., Saini, P. & De, S. Nanostructured polyaniline incorporated ultrafiltration membrane for desalination of brackish water. *Env. Sci Water Res Technol* 1, 893–904 (2015).

[45] Liu, M., Chen, C., Hu, J., Wu, X. & Wang, X. Synthesis of Magnetite/Graphene Oxide Composite and Application for Cobalt(II) Removal. *J. Phys. Chem.* C 115, 25234–25240 (2011).

[46] Zhu, L. *et al.* Mixed matrix membranes containing MIL-53(Al) for potential application in organic solvent nanofiltration. *Rsc Adv* 5, 73068–73076 (2015).

[47] Liu, F., Chung, S., Oh, G. & Seo, T. S. Three-Dimensional Graphene Oxide Nanostructure for Fast and Efficient Water-Soluble Dye Removal. *Acs Appl. Mater. Interfaces* 4, 922–927 (2012).

[48] Li, Q., Zhai, J., Zhang, W., Wang, M. & Zhou, J. Kinetic studies of adsorption of Pb(II), Cr(III) and Cu(II) from aqueous solution by sawdust and modified peanut husk. *J. Hazard. Mater.* 141, 163–167 (2007).

[49] Li, B. & Cao, H. ZnO@graphene composite with enhanced performance for the removal of dye from water. *J Mater Chem* 21, 3346–3349 (2011).

[50] Kornaros, M. & Lyberatos, G. Biological treatment of wastewaters from a dye manufacturing company using a trickling filter. *J. Hazard. Mater.* 136, 95–102 (2006).

[51] Kalantzopoulos, G. N. *et al.* Resistance to the transport of H2 through the external surface of as-made and modified silicalite-1 (MFI). *Microporous Mesoporous Mater.* 220, 290–297 (2016).

[52] Gao, W. *et al.* Engineered Graphite Oxide Materials for Application in Water Purification. *Acs Appl. Mater. Interfaces* 3, 1821–1826 (2011).

[53] Fan, W. *et al.* Hybridization of graphene sheets and carbon-coated Fe3O4 nanoparticles as a synergistic adsorbent of organic dyes. *J. Mater. Chem.* 22, 25108 (2012).

[54] Ackacha, M. A. Removal of Pb(II) from Aqueous Solution by <i>Portulaca oleracea</i> Leaves: Kinetic, Equilibrium and Thermodynamic Studies. *Am. J. Anal. Chem.* 04, 27–32 (2013).

[55] Jiao, T. *et al.* Reduced Graphene Oxide-Based Silver Nanoparticle-Containing Composite Hydrogel as Highly Efficient Dye Catalysts for Wastewater Treatment. *Sci. Reports* 5, 11873 (2015).

[56] Khin, M. M., Nair, A. S., Babu, V. J., Murugan, R. & Ramakrishna, S. A review on nanomaterials for environmental remediation. *Energy Environ. Sci.* 5, 8075 (2012).

[57] Wang, S., Sun, H., Ang, H. M. & Tadé, M. O. Adsorptive remediation of environmental pollutants using novel graphene-based nanomaterials. *Chem. Eng. J.* 226, 336–347 (2013).

[58] Shannon, M. A. *et al.* Science and technology for water purification in the coming decades. *Nature* 452, 301–310 (2008).

[59] Liu, H., Liu, Z., Yang, M. & He, Q. Surperhydrophobic polyurethane foam modified by graphene oxide. *J. Appl. Polym. Sci.* 130, 3530–3536 (2013).

[60] Tran, D. N. H., Kabiri, S., Sim, T. R. & Losic, D. Selective adsorption of oil–water mixtures using polydimethylsiloxane (PDMS)–graphene sponges. *Env. Sci Water Res Technol* 1, 298–305 (2015).

[61] Wang, S. & Peng, Y. Natural zeolites as effective adsorbents in water and wastewater treatment. *Chem. Eng. J.* 156, 11–24 (2010).

[62] Wang, H., Kang, J., Liu, H. & Qu, J. Preparation of organically functionalized silica gel as adsorbent for copper ion adsorption. *J. Environ. Sci.* 21, 1473–1479 (2009).

[63] Kasprzyk-Hordern, B. Chemistry of alumina, reactions in aqueous solution and its application in water treatment. *Adv. Colloid Interface Sci.* 110, 19–48 (2004).

[64] Hui, K., Chao, C. & Kot, S. Removal of mixed heavy metal ions in wastewater by zeolite 4A and residual products from recycled coal fly ash. *J. Hazard. Mater.* 127, 89–101 (2005).

[65] Dias, J. M., Alvim-Ferraz, M. C. M., Almeida, M. F., Rivera-Utrilla, J. & Sánchez-Polo, M. Waste materials for activated carbon preparation and its use in aqueous-phase treatment: A review. *J. Environ. Manage.* 85, 833–846 (2007).

[66] Raposo, F., De La Rubia, M. A. & Borja, R. Methylene blue number as useful indicator to evaluate the adsorptive capacity of granular activated carbon in batch mode: Influence of adsorbate/adsorbent mass ratio and particle size. *J. Hazard. Mater.* 165, 291–299 (2009).

[67] Yan, C., Wang, C., Yao, J., Zhang, L. & Liu, X. Adsorption of methylene blue on mesoporous carbons prepared using acid- and alkaline-treated zeolite X as the template. *Colloids Surfaces Physicochem. Eng. Asp.* 333, 115–119 (2009).

[68] Banat, F., S. Al-Asheh, R. Al-Ahmad & F. Bni-Khalid. Bench-scale and packed bed sorption of methylene blue using treated olive pomace and charcoal,. *Bioresour Technol* 98, 3017–3025 (2007).

[69] Namasivayam, C. & Kavitha, D. Removal of Congo Red from water by adsorption onto activated carbon prepared from coir pith, an agricultural solid waste. *Dyes Pigments* 54, 47–58 (2002).

[70] Ai, L., Li, M. & Li, L. Adsorption of Methylene Blue from Aqueous Solution with Activated Carbon/Cobalt Ferrite/Alginate Composite Beads: Kinetics, Isotherms, and Thermodynamics. *J. Chem. Eng. Data* 56, 3475–3483 (2011).

[71] Foo, K. Y. & Hameed, B. H. Detoxification of pesticide waste via activated carbon adsorption process. *J. Hazard. Mater.* 175, 1–11 (2010).

[72] Sánchez-Polo, M. & Rivera-Utrilla, J. Adsorbent-Adsorbate Interactions in the Adsorption of Cd(II) and Hg(II) on Ozonized Activated Carbons. *Environ. Sci. Technol.* 36, 3850–3854 (2002).

[73] Geim, A.K. & Novoselov, K.S.The rise and rise of graphene. *Nat. Nanotechnol.* 5, 755–755 (2010).

[74] Zhao, G., Li, J., Ren, X., Chen, C. & Wang, X. Few-Layered Graphene Oxide Nanosheets As Superior Sorbents for Heavy Metal Ion Pollution Management. *Environ. Sci. Technol.* 45, 10454–10462 (2011).

[75] Wang, J.-N. *et al.* Recent developments in superhydrophobic graphene and graphene-related materials: from preparation to potential applications. *Nanoscale* 7, 7101–7114 (2015).

[76] Sun, Y., Wu, Q. & Shi, G. Graphene based new energy materials. *Energy Environ. Sci.* 4, 1113 (2011).

[77] Schedin, F. *et al.* Detection of individual gas molecules adsorbed on graphene. *Nat. Mater.* 6, 652–655 (2007).

[78] Li, X. *et al.* Graphene Films with Large Domain Size by a Two-Step Chemical Vapor Deposition Process. *Nano Lett.* 10, 4328–4334 (2010).

[79] Zhu, J. *et al.* One-Pot Synthesis of Magnetic Graphene Nanocomposites Decorated with Core@Double-shell Nanoparticles for Fast Chromium Removal. *Environ. Sci. Technol.* 46, 977–985 (2012).

[80] Xu, J., Wang, L. & Zhu, Y. Decontamination of Bisphenol A from Aqueous Solution by Graphene Adsorption. *Langmuir* 28, 8418–8425 (2012).

[81] Ren, Y. *et al.* Graphene/d-MnO2 composite as adsorbent for the removal of nickel ions from wastewater. *Chem. Eng. J.* 175, 1–7 (2011).

[82] Luo, X. *et al.* Nanocomposites of graphene oxide-hydrated zirconium oxide for simultaneous removal of As(III) and As(V) from water. *Chem. Eng. J.* 220, 98–106 (2013).

[83] Chabot, V. *et al.* A review of graphene and graphene oxide sponge: material synthesis and applications to energy and the environment. *Energy Environ. Sci.* 7, 1564 (2014).

[84] Li, B., Cao, H. & Yin, G. Mg(OH)2@reduced graphene oxide composite for removal of dyes from water. *J. Mater. Chem.* 21, 13765 (2011).

[85] Wang, C.-F. & Lin, S.-J. Robust Superhydrophobic/Superoleophilic Sponge for Effective Continuous Absorption and Expulsion of Oil Pollutants from Water. *Acs Appl. Mater. Interfaces* 5, 8861–8864 (2013).

[86] Nguyen, D. D., Tai, N.-H., Lee, S.-B. & Kuo, W.-S. Superhydrophobic and superoleophilic properties of graphene-based sponges fabricated using a facile dip coating method. *Energy Environ. Sci.* 5, 7908 (2012).

[87] Bosma, W. J. P., Bellin, A., van der Zee, S. E. A. T. M. & Rinaldo, A. Linear equilibrium adsorbing solute transport in physically and chemically heterogeneous porous formations: 2. Numerical results. *Water Resour. Res.* 29, 4031–4043 (1993).

[88] Dutta, S. K. & Singh, D. Sorption and desorption behavior of lead in four different soils of India. *Agric. Sci.* 02, 41–48 (2011).

[89] Carrasco-Marin, F., Lopez-Ramon, M. V. & Moreno-Castilla, C. Applicability of the Dubinin-Radushkevich equation to carbon dioxide adsorption on activated carbons. *Langmuir* 9, 2758–2760 (1993).

[90] Peric, J., Trgo, M. & Vukojevic Medvidovic, N. Removal of zinc, copper and lead by natural zeolite—a comparison of adsorption isotherms. *Water Res.* 38, 1893–1899 (2004).

[91] Moreno-Castilla, C., Álvarez-Merino, M. A., López-Ramón, M. V. & Rivera-Utrilla, J. Cadmium Ion Adsorption on Different Carbon Adsorbents from Aqueous Solutions. Effect of Surface Chemistry, Pore Texture, Ionic Strength, and Dissolved Natural Organic Matter. *Langmuir* 20, 8142–8148 (2004).

[92] S, L. About the theory of so-called adsorption of soluble substances. *K Sven Vetenskapsakad Handl* 24, 1–39 (1898).

[93] Kooli, F., Yan, L., Al-Faze, R. & Al-Sehimi, A. Removal enhancement of basic blue 41 by brick waste from an aqueous solution. *Arab. J. Chem.* 8, 333–342 (2015).

[94] Mahmoodi, N. M. Equilibrium, Kinetics, and Thermodynamics of Dye Removal Using Alginate in Binary Systems. *J. Chem. Eng. Data* 56, 2802–2811 (2011).

[95] Ho, Y. . & McKay, G. Pseudo-second order model for sorption processes. *Process Biochem.* 34, 451–465 (1999).

[96] Karagoz, S., Tay, T., Ucar, S. & Erdem, M. Activated carbons from waste biomass by sulfuric acid activation and their use on methylene blue adsorption. *Bioresour. Technol.* 99, 6214–6222 (2008).

[97] Wu, X.-L., Wang, L., Chen, C.-L., Xu, A.-W. & Wang, X.-K. Water-dispersible magnetite-graphene-LDH composites for efficient arsenate removal. *J. Mater. Chem.* 21, 17353 (2011).

[98] Wang, M., Lawal, A., Stephenson, P., Sidders, J. & Ramshaw, C. Post-combustion CO2 capture with chemical absorption: A state-of-the-art review. *Chem. Eng. Res. Des.* 89, 1609–1624 (2011).

[99] Vácha, R., Slavíc?ek, P., Mucha, M., Finlayson-Pitts, B. J. & Jungwirth, P. Adsorption of Atmospherically Relevant Gases at the Air/Water Interface: Free Energy Profiles of Aqueous Solvation of N_2, O_2, O_3, OH, H_2O, HO_2, and H_2O_2. *J. Phys. Chem. A* 108, 11573–11579 (2004).

[100] Wang, S., Li, M. & Lu, Q. Filter Paper with Selective Absorption and Separation of Liquids that Differ in Surface Tension. *Acs Appl. Mater. Interfaces* 2, 677–683 (2010).

[101] Baldacchini, T., Carey, J. E., Zhou, M. & Mazur, E. Superhydrophobic Surfaces Prepared by Microstructuring of Silicon Using a Femtosecond Laser. *Langmuir* 22, 4917–4919 (2006).

[102] Wenzel, R. N. RESISTANCE OF SOLID SURFACES TO WETTING BY WATER. *Ind. Eng. Chem.* 28, 988–994 (1936).

[103] Cassie, A. B. D. & Baxter, S. Wettability of porous surfaces. *Trans. Faraday Soc.* 40, 546 (1944).

[104] Rizzuti, L. & Brucato, A. Liquid viscosity and flow rate effects on interfacial area in packed columns. *Chem. Eng. J.* 41, 49–52 (1989).

[105] Liu, Y. *et al.* Cost-Effective Reduced Graphene Oxide-Coated Polyurethane Sponge As a Highly Efficient and Reusable Oil-Absorbent. *Acs Appl. Mater. Interfaces* 5, 10018–10026 (2013).

[106] Gupta, S. S., Sreeprasad, T. S., Maliyekkal, S. M., Das, S. K. & Pradeep, T. Graphene from Sugar and its Application in Water Purification. *Acs Appl. Mater. Interfaces* 4, 4156–4163 (2012).

[107] Chen, Y., Chen, L., Bai, H. & Li, L. Graphene oxide–chitosan composite hydrogels as broad-spectrum adsorbents for water purification. *J Mater Chem* 1, 1992–2001 (2013).

[108] Wang, C. *et al.* Preparation of a graphene-based magnetic nanocomposite for the removal of an organic dye from aqueous solution. *Chem. Eng. J.* 173, 92–97 (2011).

[109] Lü, W., Wu, Y., Chen, J. & Yang, Y. Facile preparation of graphene–Fe_3O_4 nanocomposites for extraction of dye from aqueous solution. *CrystEngComm* 16, 609–615 (2014).

[110] Jiang, Y. *et al.* Magnetic chitosan–graphene oxide composite for anti-microbial and dye removal applications. *Int. J. Biol. Macromol.* 82, 702–710 (2016).

[111] Samad, Y. A., Li, Y., Schiffer, A., Alhassan, S. M. & Liao, K. Graphene Foam Developed with a Novel Two-Step Technique for Low and High Strains and Pressure-Sensing Applications. *Small* 11, 2380–2385 (2015).

[112] Sharma, P., Hussain, N., Borah, D. J. & Das, M. R. Kinetics and Adsorption Behavior of the Methyl Blue at the Graphene Oxide/Reduced Graphene Oxide Nanosheet–Water Interface: A Comparative Study. *J. Chem. Eng. Data* 58, 3477–3488 (2013).

[113] Fan, L., Luo, C., Sun, M. & Qiu, H. Synthesis of graphene oxide decorated with magnetic cyclodextrin for fast chromium removal. *J. Mater. Chem.* 22, 24577 (2012).

[114] Hao, L. *et al.* SiO2/graphene composite for highly selective adsorption of Pb(II) ion. *J. Colloid Interface Sci.* 369, 381–387 (2012).

[115] Alyüz, B. & Veli, S. Kinetics and equilibrium studies for the removal of nickel and zinc from aqueous solutions by ion exchange resins. *J. Hazard. Mater.* 167, 482–488 (2009).

[116] Rafatullah, M., Sulaiman, O., Hashim, R. & Ahmad, A. Adsorption of methylene blue on low-cost adsorbents: A review. *J. Hazard. Mater.* 177, 70–80 (2010).

[117] Li, B., Cao, H., Yin, J., Wu, Y. A. & Warner, J. H. Synthesis and separation of dyesvia Ni@reduced graphene oxide nanostructures. *J Mater Chem* 22, 1876–1883 (2012).

[118] Sun, H., Xu, Z. & Gao, C. Multifunctional, Ultra-Flyweight, Synergistically Assembled Carbon Aerogels. *Adv. Mater.* 25, 2554–2560 (2013).

[119] Tjandra, R. *et al.* Introduction of an Enhanced Binding of Reduced Graphene Oxide to Polyurethane Sponge for Oil Absorption. *Ind. Eng. Chem. Res.* 54, 3657–3663 (2015).

Chapter 3

The Application of Nanodiamond in Biotechnology and Tissue Engineering

Lucie Bacakova, Antonin Broz, Jana Liskova,
Lubica Stankova, Stepan Potocky and
Alexander Kromka

Additional information is available at the end of the chapter

http://dx.doi.org/10.5772/63549

Abstract

Diamond in the allotrope form consists of carbon atoms arranged in a cubic crystal structure covalently bonded in sp^3 hybridization. Diamond has emerged as a very promising material for various biomedical applications due to its excellent mechanical properties (hardness, low friction coefficient, good adhesiveness to the underlying substrate, good interlayer cohesion), optical properties (the ability to emit intrinsic luminescence), electrical properties (good insulator in the pristine state and semiconductor after doping), chemical resistance (low chemical reactivity and resistance to wet etching) and biocompatibility (little if any toxicity and immunogenicity). For advanced biomedical applications, diamond is promising particularly in its nanostructured forms, namely nanoparticles, nanostructured diamond films and composite scaffolds in which diamond nanoparticles are dispersed in a matrix (mainly nanodiamond-loaded nanofibrous scaffolds). This chapter summarizes both our long-term experience and that of other research groups in studies focusing on the interaction of cells (particularly bone-derived cells) with nanodiamonds as nanoparticles, thin films and composites with synthetic polymers. Their potential applications in bioimaging, biosensing, drug delivery, biomaterial coating and tissue engineering are also reviewed.

Keywords: diamond nanoparticles, nanocrystalline diamond films, nanodiamond-polymer composites, bioimaging, biosensing, drug delivery, nanodiamond cytotoxicity, biomaterial coating, tissue engineering

1. Introduction

The approaching nano- and biotechnological era expects the implementation of "smart" and "functional" materials, which will be used not only as a mechanically passive substrates, but also as active devices (i.e., electronic or optical devices and sensors, micro-electro-mechanical systems, smart drug delivery systems, bioactive substrates for the cell adhesion and growth, etc.).

Diamond represents a class of perspective multifunctional materials with morphological, chemical, optical and electronic properties tailorable on demand for specific needs, especially for life science, tissue engineering or regenerative medicine.

The bonds between the carbon atoms in the diamond lattice are covalent and predominantly of sp^3 hybridization [1, 2]. The diamond lattice of covalently bonded carbon atoms endows the diamond with extraordinary combination of intrinsic properties, particularly high hardness and thermal conductivity [3, 4]. The fracture toughness of the diamond has been measured to be 2 MPa $m^{1/2}$, which is a relatively high value compared to other gemstones or other optical materials [5]. Diamond toughness is also dependent on the crystallographic plane on which the fracture force is expressed. The lowest cleavage energies were measured for the <111> plane [6]. The combination of the highest Young's modulus, a high fracture toughness, a low friction coefficient and a low thermal expansion predetermines it as a material for protective layers and coatings. From optical point of view, diamond offers a unique combination of transparency in most of the ultraviolet, visible and infrared regions (from 227 nm to far IR). The refractive index of diamond is 2.41–2.44 (656–486 nm). As a semiconductor, diamond has an excellent combination of properties (except of electron mobility) [7, 8]. Other remarkable properties of diamond include a high wear resistance [9–11], pressure resistance [12], strong adhesion to the underlying substrate, i.e., when deposited in the form of films [10, 13, 14], a low friction coefficient [15] and high chemical stability (e.g., low reactivity and resistance to wet etching [2]). These properties can be further improved by various techniques of material engineering, e.g., by modulating the synthesis and structure of diamond and by various combinations of diamond with other elements and compounds. For example, hardness, fracture toughness, Young's modulus and wear resistance of diamond can be markedly enhanced by the preparation of diamond in the form of aggregated diamond nanorods [9], its hardness and thermal stability can be increased by nanotwinning with cubic boron nitride [4] and the adhesion of diamond films can be improved by laser treatment of the underlying substrate [14].

The optical properties of the diamond are also excellent. Pure diamond transmits visible light and appears as a clear colorless crystal. However, diamond is capable of high optical dispersion (i.e., the ability to disperse light of different colors). In addition, although the diamond lattice is very strong and rigid, it can contain defects or be contaminated with foreign atoms (N, B, H, Ni, Co, Cr, Si), e.g., during the growth of diamond lattice. These elements can also be introduced into this lattice by ion implantation. The lattice defects and the presence of foreign atoms are responsible for the various colors of diamond, e.g., yellow and orange (nitrogen), brown (nitrogen and lattice defects), blue (boron), green (trace amounts of nickel or radiation exposure), gray (the presence of boron or unsaturated forms of carbon), black (inclusions of

http://dx.doi.org/10.5772/63549

graphite and iron) or also purple, pink and red (due to changes in the electron structure by plastic deformation during traveling of diamonds from the Earth's mantle to its surface *via* magma) [16]. The defects and the presence of foreign atoms in the diamond lattice also produce the intrinsic luminescence (fluorescence) of diamond [2, 17–19]. The electrical properties of diamond can be tuned by dopants. Pure diamonds can act as excellent electrical insulators but after chemical doping, e.g., the incorporation of boron, they are converted into a p-type semiconductor [20] or superconductor [21]. Boron is by far the most widely used dopant but other dopants are also possible [22].

For all these outstanding properties, diamond is an attractive material for technical and industrial application, e.g., for polishing, machining, cutting and drilling tools. These tools are also useful for biomedical applications, e.g., for polishing biomaterials with diamond paste or cutting them with a diamond saw [23, 24], for bone resection or craniotomy using a diamond threadwire saw [25, 26] or for dental drilling in stomatology [27]. Other important fields for diamond application are electronics, optics or the jewelry trade.

This chapter will concentrate on newly explored applications of diamond in biotechnologies and life sciences, such as bioimaging, biosensing, tissue engineering, controlled drug and gene delivery and for detecting and capturing various biomolecules and coating body implants. For some of these uses, diamond is attractive mainly in its nanosized or nanostructured form, namely free nanoparticles (1D), planar films from nanodiamonds (2D), and composites of diamond nanoparticles (DNPs) and other molecules, particularly polymers in the form of 3D scaffolds, as shown in **Figure 1**.

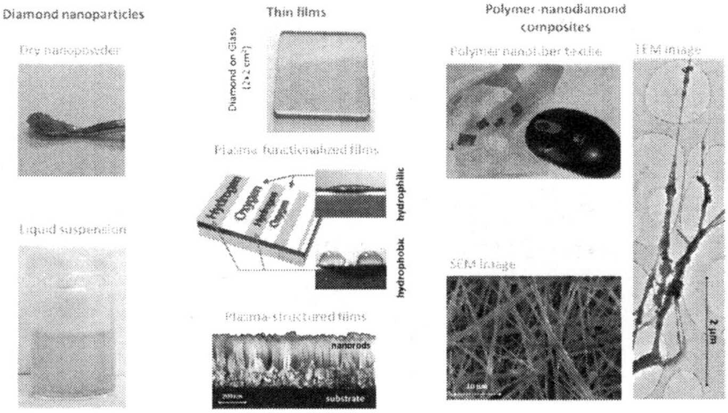

Figure 1 Schematic view of three diamond material forms representing free nanoparticles (1D), planar nanodiamond films (2D) and composite 3D scaffolds.

Our earlier studies included in this chapter were focused mainly on nanocrystalline diamond (NCD) films as substrates for the adhesion, growth and osteogenic differentiation of human bone-derived cells in the form of commercially available cell lines, namely human osteoblast-like MG 63 cells [20, 28–33] or Saos-2 cells [34–41], primary osteoblasts [42] and bone marrow mesenchymal stem cells (MSCs) [34, 42, 43]. The cell behavior on NCD films was further modulated by the size of the surface irregularities, measured by the root mean square (RMS) roughness parameter [34, 35], by the shape of these irregularities (nanocones and nanorods [36, 37], by termination of the NCD films with oxygen or hydrogen [38–42] and by doping of these films with boron [20, 32, 33]. NCD films micropatterned with H-terminated and O-terminated domains were also used for construction of biosensors [44]. Some of our studies were also dedicated to the construction of polymeric nanofibrous scaffolds reinforced with DNPs as potential scaffolds for bone tissue engineering. All the mentioned studies are summarized in **Table 1**.

References	Topic	Cell model
Bacakova *et al.* [28]	Cell adhesion and proliferation on nanostructured and hierarchically organized submicron- and nanostructured NCD films	MG 63
Grausova *et al.* [29]	Cell adhesion and proliferation on nanostructured and hierarchically organized submicron- and nanostructured NCD films	MG 63, endothelial CPAE cells
Grausova *et al.* [30]	Adhesion, proliferation, viability, mitochondrial activity and osteogenic differentiation of cells on nanostructured and hierarchically organized submicron- and nanostructured NCD films	MG 63
Grausova *et al.* [31]	Adhesion, osteogenic cell differentiation and immune activation of cells on nanostructured and hierarchically organized submicron- and nanostructured NCD films	MG 63
Broz *et al.* [34]	Regulation of the cell adhesion by the surface roughness of NCD films	Saos-2 MSC
Kalbacova *et al.* [35]	Regulation of the osteogenic cell differentiation by the surface roughness of NCD films	Saos-2
Babchenko *et al.* [37]	Regulation of the cell adhesion by the shape of surface irregularities of the NCD films	Saos-2
Kalbacova *et al.* [36]	Regulation of the cell adhesion, activity of focal adhesion kinase and osteogenic cell differentiation by the shape of surface irregularities of the NCD films	Saos-2
Rezek *et al.* [38]	Adhesion and growth of cells on NCD films patterned with O-terminated and H-terminated microdomains	Saos-2
Ukraintsev *et al.* [40]	Adsorption of fetal bovine serum proteins and cell adhesion on NCD films patterned with O-terminated and H-terminated microdomains	Saos-2
Rezek *et al.* [39]	Adsorption of fibronectin and cell adhesion on NCD films patterned with O-terminated and H-terminated microdomains	Saos-2

References	Topic	Cell model
Liskova *et al.* [41]	Adhesion, growth and osteogenic differentiation of cells on O- and H-terminated NCD films	Saos-2
Liskova *et al.* [42]	Adhesion, growth and osteogenic differentiation of cells on O- and H-terminated NCD films	MSC, primary osteoblasts
Izak *et al.* [44]	Application of NCD films patterned with O- and H-terminated microdomains for construction of impedance-based sensor for real-time monitoring of cell growth	MG 63
Kopecek *et al.* [33]	Adhesion, growth and osteogenic differentiation of cells on boron-doped NCD films	MG 63
Kromka *et al.* [20]	Adhesion, growth and osteogenic differentiation of cells on boron-doped NCD films	MG 63
Grausova *et al.* [32]	Adhesion, growth and osteogenic differentiation of cells on boron-doped NCD films	MG 63
Parizek *et al.* [45]	Cell adhesion and growth on nanofibrous PLGA-nanodiamond scaffolds	MG 63
Brady *et al.* [43]	Cell adhesion and growth on nanofibrous PLGA-nanodiamond scaffolds	MSC
Bacakova *et al.* [46]	Cell adhesion and growth on nanofibrous PLLA-nanodiamond scaffolds	MG 63, Saos-2

Table 1. Summarization of our earlier studies on cell performance on NCD films and polymeric nanofibrous scaffolds loaded with diamond nanoparticles.

2. Diamond nanoparticles

Individual DNPs can be advantageously used for drug and gene delivery, bioimaging technologies and biosensor construction. For these purposes, it is necessary to attach various atoms or molecules on the DNP surface, to achieve good dispersiveness of the DNPs in aqueous solutions, such as drug vehicles, buffers, physiological solution, cell culture media or body fluids and also to add chemical functionality to the DNPs in order to enable the uptake of DNPs by cells. DNPs have a high surface/volume ratio, i.e., they have a relatively large surface, on which not only atoms and small molecules, but also macromolecules (e.g., various drugs, nucleic acids and proteins) can be attached by physi- or chemisorption (for a review, see [47]). However, this attachment can be limited, for example, due to DNP hydrophobicity and tendency to aggregate in aqueous solutions. Thus, it is necessary to engineer the surface of DNPs for their specific applications [2]. The attachment of synthetic and biological molecules to the nanodiamond surface and the dispersion of DNPs in the aqueous media can be improved by various types of functionalization of the nanodiamond surface, e.g., by oxygen-containing chemical functional groups, such as –OH, –COOH [48–50], amine groups [48, 51] and by various biomolecules, such as lysine [52], biotin [53], thionine, trimethylamine [49], polyglycerol [54, 55] and RGD-containing oligopeptides, i.e., ligands for adhesion receptors on the cell surface [54, 56]. This functionalization can also improve the penetration of DNPs into cells (i.e., by crossing cytoplasmic and nuclear membranes) or by their internalization through clathrin-mediated endocytosis [57–59]. Functionalization of the DNPs with thiol groups (–SH) enables their conjugation with gold and silver nanoparticles [60, 61]. Such complexes are promising for photothermal therapy against tumors [60].

2.1. Controlled drug and gene delivery

Nanodiamond-based drug delivery has been developed particularly for advanced tumor therapies, e.g., the treatment of multidrug-chemoresistant leukemia. In experiments *in vitro*, daunorubicin conjugated with DNPs was efficient against a multidrug-resistant K562 human myelogenous leukemia cell line, which was able to overcome treatment with daunorubicin alone [62]. Other anticancer drugs, which were successfully conjugated with DNPs and delivered to cells, were 10-hydroxycamptothecin [63], polymyxin [64] or doxorubicin, an apoptosis-inducing drug widely used in chemotherapy [54, 65]; for a review, see [1, 66]. A nanodiamond-mediated doxorubicin delivery system was designed to inhibit the lung metastasis of breast cancer [67] and also to treat malignant brain gliomas [68]. The nanodiamond-doxorubicin complexes were more significantly efficient in killing glioma cells *in vitro* and also *in vivo* in rats than the uncomplexed drug [68].

Other therapeutic molecules which can be delivered into cells through DNPs include cell growth and differentiation factors, antibodies, vaccines and anti-inflammatory drugs. For example, basic fibroblast growth factor (bFGF) and bone morphogenetic protein-2 (BMP-2) were loaded onto DNPs by physisorption, forming a stable colloidal solution. This solution enabled the release of both factors in slightly acidic conditions, induced the proliferation and osteogenic differentiation of osteoblast progenitor cells and was injectable. Thus, this solution was an effective alternative to the currently used delivery of bFGF and BMP-2 to the surgical site by the implantation of bulky collagen sponges [69]. Carboxylated DNPs conjugated with growth hormone were tested for a specific targeting growth hormone receptor in cancer cells and potential anticancer therapy [70].

DNPs can also serve as a stable delivery platform for therapeutic antibodies, as revealed by studies on the transforming growth factor β (TGF-β) antibody. The complexes of this antibody with DNPs were stable in water, but under physiological conditions simulated in serum-supplemented cell culture, the release of the active antibody was detected [71]. An interesting newly developed application of DNPs is delivering vaccines through the skin for painless and efficient immunization. These nanoparticles could improve limited drug delivery through the stratum corneum, which has a very dense structure and only allows the penetration of small molecules with a molecular weight of below 500 Da (for a review, see [72]). Fluorescent nanodiamonds functionalized by hyperbranched polyglycerol also proved to be promising carriers for delivering aluminum oxyhydroxide, i.e., a compound widely used as an immunologic adjuvant of vaccines [73].

For localized drug delivery, DNPs can be incorporated into various materials, e.g., films for potential implant coating [74]. For example, aqueous dispersible detonation nanodiamonds were assembled into a closely packed multilayer nanofilm with positively charged poly-L-lysine *via* the layer-by-layer deposition technique, and integrated with drugs suppressing the release of inflammatory cytokines (tumor necrosis factor-alpha, interleukin-6) and nitric oxide synthase induced by lipopolysaccharides [65].

DNPs can also be effectively used for the proteolysis and digestion of glycopeptides. Trypsin and peptide-N-glycosidase F, immobilized on detonation nanodiamond, were more efficient than the free enzymes or commercially available beads immobilized with trypsin [75].

As for gene delivery, DNPs can act as excellent nonviral vectors for binding and transferring plasmid DNA and small interfering RNA (siRNA), particularly after functionalization, e.g., with –OH groups [49] or lysine [52]. DNPs, rendered cationic by coating with polyethyleneimine or by termination with hydrogen in plasma, efficiently delivered siRNA into Ewing sarcoma cells and blocked the expression of EWS/FLI-1 oncogene in these cells. In addition, the association of EWS/FLI-1 silencing by the siRNA/nanodiamond complex with a vincristine treatment potentiated the cytotoxic effect of this chemotherapeutic drug [59]. Similarly, DNPs conjugated with RGD oligopeptides and siRNA for vascular endothelial growth factor (VEGF) can be used in antiangiogenic gene therapy to inhibit tumor growth *via* the downregulation of the VEGF expression [56]. Last but not least, an array of DNPs in the form of nanoneedles facilitated the delivery of double-strand DNAs (dsDNA90) into cells to activate the pathway involving the stimulator of interferon genes (STING). In addition, this array was successfully utilized to isolate the transcriptional factor, NF-kB, from intracellular regions without damaging the cells, upon STING activation [76].

2.2. Bioimaging

Due to stable and controllable photoluminescence, DNPs are also promising for advanced bioimaging techniques. Like the various colors of diamonds, mentioned above, this photoluminescence is based on crystallographic defects and particularly on optical centers created by the incorporation of foreign atoms (N, Si or Cr) into the diamond lattice. DNPs with a nitrogen-vacancy (NV) color center were detectable in living cells *in vitro,* and also after application into chicken embryos *in vivo,* using a standard confocal microscope [19]. In addition, DNPs with NV color centers offer great potential for advanced imaging techniques, such as optical super resolution nanoscopy, magneto-optical (spin-assisted) subwavelength localization and imaging, single molecule spin imaging or nanoscale imaging of biomagnetic fields [17]. Photoluminescent nanodiamonds of size less than 50 nm were used to investigate the mechanism of their uptake by living cells. By selectively blocking different uptake processes, it was shown that DNPs enter cells mainly by clathrin-mediated endocytosis. Intracellularly, the largest nanoparticles and aggregates were localized in vesicles, while the smallest particles appeared free in the cytosol [58]. Fluorescent DNPs were also used for tracking the transplanted lung stem cells in mice with naphthalene-induced lung injury, and these stem cells were found at terminal bronchioles of the lungs for 7 days after intravenous transplantation [77].

The optical activity of the luminescent color centers in DNPs depends on their proximity to the nanoparticle surface and its surface termination [2]. The photoluminescence can also be modulated by changes in the environment surrounding the DNPs. For example, coating the DNPs with a polymer film resulted in a reduction of the lifetime of NV centers and an average enhancement in their emission rate [78]. For photoacoustic imaging in biological tissues, nanodiamonds with very high absorption in the near-infrared range can be created by irradiation with the He^+ ion beam [79]. The fluorescence of DNPs can also be induced by surface passivation of the DNPs with bis(3-aminopropyl) terminated poly(ethylene glycol) [60]. A nanodiamond-polyglycerol-gadolinium(II) conjugate was designed and prepared as a novel nanodiamond-based magnetic resonance contrast agent dispersible in physiological media [55]. For combined bioimaging and drug delivery, photoluminescent DNPs were coated with a porous SiO_2 shell [80].

2.3. Biosensing

DNPs with an NV color center are also emerging tools for nanoscale sensing, e.g., sensing molecular fluctuations and temperatures in live cellular environments, detecting and measuring magnetic and electric fields, and measuring ion concentrations and cell membrane potentials [17, 18]. DNPs immobilized onto a gold electrode by direct adsorption were used as a biosensor for determining the concentration of glucose and lactate [81]. Arrays of DNPs in the form of nanoneedles were applied for intracellular sensing, e.g., of intracellular signaling through NF-kB, translocated from the cell cytoplasm to the nucleus region [76]. Other sensoric applications of nanodiamonds require the creation of continuous diamond films, and they are discussed in Section 3 (Nanostructured diamond films).

2.4. The potential cytotoxicity of diamond nanoparticles

For all the applications mentioned above, it is necessary to use nontoxic nanoparticles. DNPs are generally considered as materials with little, if any cytotoxicity. These nanoparticles dispersed in the cell culture media did not affect the adhesion, growth, viability and differentiation of various cell types (for a review, see [82, 83]). For example, the addition of DNPs in the culture medium did not affect the adhesion, growth and adipogenic or osteogenic differentiation of human adipose tissue-derived stem cells [84]. Also the labeling of lung stem cells with fluorescent DNPs *in vivo* in mice did not affect the ability of these cells to grow and to differentiate into type I and type II pneumocytes [77].

In spite of these encouraging findings, the number of reports on nanodiamond cytotoxicity is increasing. The first report demonstrated the damage and destruction of human erythrocytes and leucocytes by DNPs [85, 86]. For low and moderate concentrations, DNPs did not negatively influence the viability of isolated inflammatory neutrophils and even enhanced their response to bacterial agents. However, the viability and antibacterial reaction of these cells was suppressed at high concentrations of DNPs (≥ 1 g/l; [87]). Similar reactions of neutrophils were also observed *in vivo* after intravenous and intraperitoneal administration of DNPs into rats [88]. DNPs evoked significant activation of human platelets, and when administered intravenously in mice, they induced widespread pulmonary thromboembolism [89]. DNPs had antiangiogenic effects, exerted by the downregulation of the gene and protein expression of the proangiogenic bFGF [90]. DNPs also induced apoptosis and necrosis of vascular endothelial cells [91, 92] and caused DNA damage of mouse embryonic stem cells [93] and human glioblastoma U87 cells [94]. Moreover, DNPs decreased the proliferation and viability of various cell types, e.g., human cervical carcinoma HeLa cells [95], human osteoblast-like MG 63 cells and primary rat MSCs [96].

The adverse effects of DNPs have usually been attributed to the oxidative stress generated by these particles [92, 95, 97, 98]. DNPs created by detonation synthesis affected the cellular content of glutathione and the activities of the main antioxidant enzymes (superoxide dismutase, catalase, glutathione peroxidase, glutathione reductase and glutathione S-transferase) in human umbilical vein endothelial cells *in vitro* [98]. Furthermore, after intravenous and intraperitoneal administration into rats *in vivo*, DNPs increased the activity of superoxide dismutase and decreased the activity of glutathione reductase and glutathione peroxidase within erythrocytes [88]. In addition, DNPs increased the adsorption and delivery of a large amount of sodium ions into cells, which induced osmotic stresses, cell swelling, increase in the intracellular levels of calcium and severe cellular damage [97].

The cytotoxicity of DNPs depends on their origin, size, surface functionalization, tendency to form aggregates and the presence of impurities. In studies reporting the adverse effects of DNPs, these nanoparticles were often prepared by detonation [85, 86, 90, 93, 96, 97], while in studies indicating the nontoxic effects of DNPs on cell behavior, the DNPs were prepared by non-detonation methods, e.g., milling diamond microcrystals [73, 99]. The detonation DNPs, also called ultrananocrystalline detonation diamond (UDD), are generally of smaller size (in nanometers) than non-detonation DNPs (tens of nanometers), and this small size has been considered as the main determinant of nanoparticle cytotoxicity [91]. The detonation diamonds often contain impurities, such as other carbon allotropes [53], iridium [100], carbon soot and oxides and carbides, including those of iron, chromium, silicon, calcium, copper, potassium, titanium and sulfur (for a review, see [1]). In a study by Keremidarska et al. [96], the increased cytotoxicity of DNPs was associated with their smaller size and presence of impurities, particularly other carbon allotropes. Thus, for biomedical studies, detonation nanodiamonds need to be extensively purified. For the removal of nondiamond carbon, the following chemicals can be used: ozone-enriched air, liquid oxidants such as HNO_3, $HClO_4$ or different acid mixture under pressure; metal impurities can be removed by treatment with HCl, H_2SO_4 and its mixtures with HNO_3 or potassium dichromate (for a review, see [1, 96]).

The functionalization influence of detonation DNPs on their biocompatibility is controversial. Detonation nanodiamonds covered by cobalt ions showed a protective effect against contact dermatitis in mini-pigs, induced by local exposure of the animal skin to these ions [101]. Detonation nanodiamonds with hyperbranched polyglycerol coating showed good cytocompatibility with A549 cancer cells and U937 macrophages [54]. Detonation nanodiamonds functionalized with thiol groups and conjugated with gold nanoparticles did not show significant cytotoxicity to a human lung carcinoma cell line [61]. On the other hand, detonation DNPs terminated with oxygen showed higher cytotoxicity to mouse embryonic stem cells than pristine unmodified DNPs, although this cytotoxicity was lower than in multiwalled carbon nanotubes [93].

In our studies, we compared the potential cytotoxicity of DNPs from three different origins. DNPs fabricated in the Nano Carbon Research Institute (Japan) under the product name NanoAmando were prepared by a detonation method and were hydrogen-terminated. A part of these DNPs was oxygen-terminated by annealing at 450°C in air. The size of both unmodified and O-terminated detonation DNPs was only about 5 nm. The DNPs, which were acquired from Microdiamant AG (Switzerland) named MSY (Monocrystalline SYnthetic), were prepared using a high pressure high temperature (HPHT) method and were oxygen-terminated due to postproduction acid treatment by the producer. These DNPs were not further modified and were provided in several sizes ranging from 18 to 210 nm. The DNPs acquired from Adámas Nanotechnologies, Inc. (Raleigh, U.S.A.) were prepared using the HPHT method and were already provided in a water colloid with a concentration of 1000 mg/l. The size of these nanoparticles was 40 nm. The Adámas DNPs had intrinsic red fluorescence based on NV defects in their diamond lattice, which was further activated by annealing in a vacuum at 800°C.

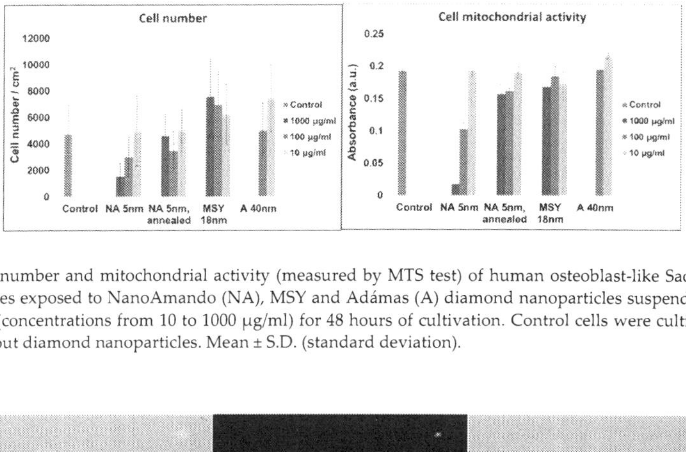

Figure 2. The number and mitochondrial activity (measured by MTS test) of human osteoblast-like Saos-2 cells in 3-day-old cultures exposed to NanoAmando (NA), MSY and Adámas (A) diamond nanoparticles suspended in cell culture medium (concentrations from 10 to 1000 µg/ml) for 48 hours of cultivation. Control cells were cultivated in pure medium without diamond nanoparticles. Mean ± S.D. (standard deviation).

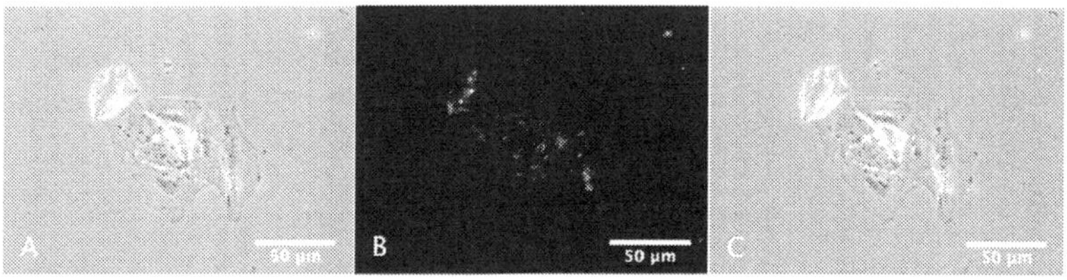

Figure 3. Micrographs of cells cultivated for 48 hours with Adámas DNPs Olympus IX 71, obj. 40×. Phase contrast image (A), red fluorescence of Adámas DNPs in grayscale (B), and merged images A and B (C).

The DNPs were dispersed by ultrasonication in water to obtain a colloid with a DNP concentration of 10,000 µg/ml, and added to 24-hour-old cultures of human osteoblast-like Saos-2 cells in concentrations of 10, 100 or 1000 µg/ml of the culture medium. After 48 hours, the cell number was evaluated by cell counting on recorded microphotographs using Image J software

and correlated with the activity of mitochondrial enzymes measured by a commercially available MTS test (Promega). The lowest cell numbers and mitochondrial activity were obtained in cultures with unmodified NanoAmando DNPs, and these values decreased with increasing nanoparticle concentration (**Figure 2**). The cell numbers slightly improved in cultures with annealed NanoAmando DNPs, but they did not reach the average values obtained in MSY and Adámas DNPs. Only the mitochondrial activity of cells cultured with annealed NanoAmando DNPs reached similar values to the cells exposed to MSY and Adámas DNPs. The annealing of DNPs at a high temperature probably had a similar effect to the chemical purification of DNPs in oxidizing agents. The fluorescence of Adámas DNPs was detected inside the fixed cells after 2 days of cultivation using the standard epifluorescence microscope Olympus IX 71 (**Figure 3**).

3. Nanostructured diamond films

3.1. Diamond films as substrates for cell adhesion and growth

Our earlier studies and studies by other authors showed that NCD films provided excellent substrates for the adhesion, growth and phenotypic maturation of various cell types, such as neuronal and glial cells [102–104], epithelial cells [105], vascular endothelial cells [29], fibroblasts [106] and particularly bone-derived cells (for a review, see [82, 83, 107]). The latter cells included commercially available lines of osteoblast-like cells (MG 63 [20, 28–33], Saos-2 [30, 31, 35, 41, 108]), primary human osteoblasts [42] and human bone marrow MSCs [42, 108, 109].

The positive effects of NCD films on cell performance can be explained by their nanostructure, i.e., the presence of irregularities at the nanoscale level on their surface (e.g., an average roughness R_a parameter equal or less than 100 nm). On these surfaces, the cell adhesion-mediating proteins present in biological fluids (e.g., fibronectin and vitronectin in the serum supplement of the culture media) are adsorbed in a favorable geometrical conformation enabling the exposure of specific adhesion sites in these proteins (e.g., RGD motifs) to the cell adhesion receptors, i.e., integrins. In addition, it is believed that the nanostructured surfaces preferentially adsorb vitronectin (due to its relatively small and linear molecule), which is favorably recognized by osteoblasts through its KRSR amino acid sequence [110, 111] (for a review, see [107, 112]). Thus, the NCD films are promising particularly for the surface coating of bone and dental implants. In accordance with this concept, submicron crystalline diamond films (grain sizes 200–1000 nm) and particularly microcrystalline diamond films (grain size up to 2 μm) supported the adhesion and spreading of osteoblasts to a lesser extent than NCD films [113, 114]. Also in our earlier study, the number of human osteoblast-like Saos-2 cells adhering to NCD films deposited on silicone substrates of various roughness (RMS of 20, 270 and 500 nm) decreased with the increasing substrate roughness [34]. In addition, the NCD films of RMS of 20 nm and 270 nm supported better the osteogenic differentiation of Saos-2 cells than the surfaces of a RMS of 500 nm, as manifested by a higher activity of alkaline phosphatase and a higher cellular content of calcium and phosphates [35]. Similarly, on NCD films with hierarchically-organized submicron and nanoscale surface roughness (RMS of 301.0 nm and 7.6

nm, respectively), the MG 63 cells adhered in lower initial numbers than on nanorough only surfaces (RMS 8.2 nm), although the subsequent cell proliferation was higher on the hierarchically-organized surfaces [28].

Not only the size but also the shape of the surface irregularities on the diamond films is an important factor regulating the cell behavior. For example, in our earlier studies, two different types of nanoscale features, namely nanocones and nanorods, were prepared by plasma etching of nanodiamond films using Au and Ni masks, respectively. As revealed by immunofluorescence staining of vinculin, Saos-2 cells cultivated on relatively short and broad nanocones (height 5–100 nm, diameter up to 80 nm) formed less numerous but large focal adhesions, while the cells cultivated on relatively high and thin nanorods (height 120–200 nm, diameter 20–40 nm) formed more numerous, but very thin and fine focal adhesions [37]. The large focal adhesions on nanocones were associated with stronger cell adhesion, increased activation of focal adhesion kinase (FAK), and were better predestined for osteogenic cell differentiation [36].

In contrast to free DNPs, to the best of our knowledge, no study reported the cytotoxicity of NCD films. This may be due to the fact that NCD films are compact, chemically and mechanically stable and their surface is much better defined/controlled than the surface of DNPs. Moreover, planar NCD films do not interfere and/or penetrate into the cells. The surface termination of NanoAmando DNPs controlled by high-temperature treatment was shown as a crucial parameter for their cytotoxicity (see Section 3.4). NCD films were not cytotoxic even if NanoAmando DNPs were used for the substrate seeding prior to NCD deposition.

NCD films have usually been synthesized from methane or fullerene C_{60} precursors using the microwave plasma-enhanced chemical vapor deposition (MW PECVD) method. NCD films have been deposited on various substrates such as the silicon, glass or metals currently used in orthopedics and stomatology, e.g., Ti [115] and Ti-6Al-4V [116] (for a review, see [82, 83]). On these substrates, the NCD films behave as well-adhering, highly cohesive and mechanically and chemically resistant coatings. Not only were their potential adverse effects suppressed, but no particles were released from the compact films, which in addition behaved as a passivation barrier for particles and ions from the underlying materials. For example, silicon wafers, which are currently being used as experimental substrates for NCD deposition and behaved as cytotoxic for human osteoblast-like MG 63 cells and vascular endothelial CPAE cells, were rendered highly compatible with these cells by their continuous and hermetic encapsulation with NCD films [29].

Cell performance on NCD films can be further modulated by their functionalization with various atoms, chemical functional groups or biomolecules. For example, the adhesion, growth and osteogenic differentiation of bone-derived cells were better on O-terminated than on H-terminated NCD films. Human osteoblast-like Saos-2 cells, primary human osteoblasts and bone marrow MSCs in cultures on O-terminated NCD films showed better-developed focal adhesion plaques (**Figure 4**), reached higher cell population densities and produced more collagen I and alkaline phosphatase (i.e., an early and middle marker of osteogenic cell differentiation, respectively). The cells on O-terminated surfaces also showed a higher activity of alkaline phosphatase, i.e., an enzyme participating in bone matrix mineralization, and

http://dx.doi.org/10.5772/63549

accordingly with it, these cells produced more calcium in their extracellular matrix (ECM) [41, 42]. Furthermore, human dental stem cells cultured on O-terminated diamond films deposited more ECM than on H-terminated films, and this matrix contained higher levels of calcium, oxygen and phosphorus [117]. Similar behavior has also been observed in human renal epithelial HK-2 cells cultured on borosilicate glass with NCD films. On the films terminated with O, these cells adhered and grew better than on H-terminated NCD films and uncoated borosilicate glass [105]. When NCD films were constructed as micropatterned films, i.e., with domains terminated with O or H, e.g., in the form of stripes of 30–200 μm in width, the cells adhered and grew preferentially on the O-terminated domains [38, 117].

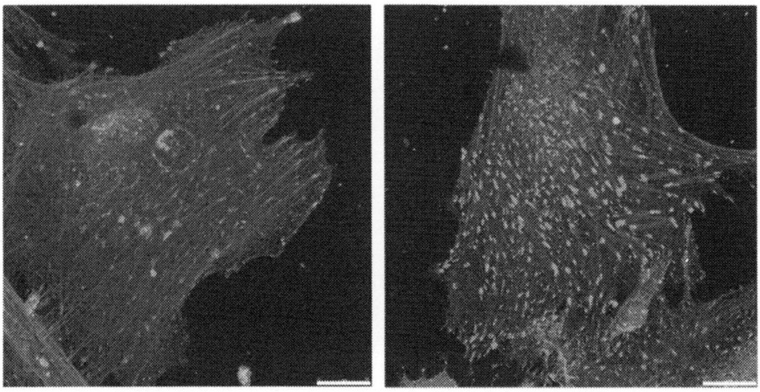

Figure 4. The immunofluorescence of paxillin (green), a protein of focal adhesion plaques, in primary human osteoblasts on day 7 after seeding on H-terminated (Diam. H) and O-terminated (Diam. O) NCD films. Actin cytoskeleton is stained with Phalloidin-Texas Red, cell nuclei with Hoechst #33342 (blue). Bar = 25 μm.

The positive influence of O-terminated diamond films on cell behavior has been explained by the higher polar component of the free surface energy and higher wettability of these films [118] (for a review, see [41]). It is known that wettable surfaces (similar to the nanostructured surfaces mentioned above) promote the adsorption of cell adhesion-mediating molecules in a bioactive and flexible geometrical conformation, and their effective recognition and binding by the cell adhesion receptors (for a review, see [112]). Our earlier study showed that the fetal bovine serum (FBS), i.e., an important supplement of standard cell culture media and an important source of cell adhesion-mediating proteins (e.g., vitronectin, fibronectin), formed more compact and better-adherent films on O-terminated NCD surfaces than on H-terminated NCD, where the FBS layer tended to peel from the material surface [40]. The following experiments with preadsorption of NCD films with O-terminated and H-terminated microdomains with FBS proteins revealed that the preferential growth of Saos-2 cells on O-terminated domains is mediated by fibronectin [39].

Positive effects on surface polarity, wettability and cell adhesion were also observed in NCD films terminated with amine groups, which upregulated the adhesion of human femur osteoblasts [115, 118], or supported the adhesion of dorsal root ganglion neurons and Schwann cells isolated from Wistar rats and NG108-15 neuroblastoma-glioma hybrid cells [102].

However, in the case of the neural stem cells derived from mouse embryos, both the O-terminated and H-terminated NCD surfaces effectively supported well the cell proliferation, and this proliferation was even slightly better on H-terminated surfaces. Moreover, H-terminated NCD films were able to spontaneously induce (i.e., without the presence of differentiation factors in the culture medium) differentiation of the stem cells into neurons, while on O-terminated NCD films, the cells preferentially differentiated into oligodendrocytes [103]. The neuronal cell differentiation on H-terminated NCD films was explained by the high accumulation of fibronectin on these films, which then activated a signaling pathway involving β_1-integrin adhesion receptors, FAK and mitogen-activated protein kinase/extracellular signaling-regulated kinase1/2 (MAPK/Erk1/2) [104]. Thus, the H-terminated surfaces seemed to have a greater potential to promote the regeneration of neurons in a damaged central nervous system (which is considered as impossible in human patients) than O-terminated surfaces.

Thus, the results of cell performance on H-terminated and O-terminated NCD films, obtained by various authors, are controversial. This disproportion can be attributed to different NCD qualities (the presence of non-diamond carbon phases, different grain size, etc.), the different reactivity of various cell types (neural, epithelial, osteogenic) or different cultivation conditions. For example, the studies by Chen et al. [103, 104] were performed using a medium with a low concentration of serum (only 2% of FBS), while the experiments with osteogenic cells were generally carried out in standard cell culture media supplemented with 10–15% of FBS. When osteogenic cells were cultured in a serum-free medium, no differences in cell adhesion on O- and H-terminated NCD domains were observed [38]. Another explanation could be the relatively high hydrophilicity of O-terminated surfaces (water drop contact angle ~20° *vs.* ~78° for H-terminated surfaces) in studies by Chen et al. [103, 104]. It is generally known that cell adhesion and growth is optimal on moderately hydrophilic surfaces (for a review, see [112]). On highly hydrophilic surfaces, the adsorption of cell adhesion-mediating proteins is weak and unstable, which hamper or disable cell adhesion. For example, extremely hydrophilic O-terminated nanostructured diamond surfaces (contact angle ~2°) almost completely resisted the adhesion of human bone marrow MSCs, whereas less hydrophilic H-terminated nanodiamond surfaces (contact angle ~86°) gave good support for the adhesion and growth of these cells [109]. Also in our studies, human osteoblast-like MG 63 cells on O-terminated NCD films (contact angle about 20–35°) showed a higher tendency for spontaneous detachment at later culture intervals, when the surfaces had to hold a large number of cells [29–31] than the cells on H-terminated NCD films (contact angle about 85–90° [33], (for a review, see [112].

Another important modification of NCD films influencing cell behavior is the boron doping of these films. Lower concentrations of boron (approximately 100–1000 ppm of B) supported the proliferation and early osteogenic differentiation of human osteoblast-like MG 63 cells, which was manifested by an increased cell number and increased concentration of collagen I and alkaline phosphatase in the cells. Higher concentrations of boron (6700 ppm of B) enhanced cell adhesion and osteogenic cell differentiation, manifested by increased concentrations of vinculin, a focal adhesion protein, and osteocalcin, a calcium-binding ECM glycoprotein. These effects were probably due to the increased electrical conductivity of NCD films after boron doping [32]. It is known that electroactive materials are able to enhance the adhesion, growth and phenotypic maturation on various cell types, even without active stimulation with an electrical current (for a review, see [83, 112]).

Also, composite apatite-nanodiamond coatings were shown to improve the performance of bone cells. When electrodeposited on stainless steel, these coatings markedly enhanced the attachment, spreading and formation of focal adhesion plaques in osteoblast-like MG 63 cells, compared to the pure stainless steel and apatite coatings without nanodiamond, which was mediated by an increased adsorption of fibronectin on the composite coatings and its deposition by the cells [119].

3.2. Diamond films as substrates for biosensing and localized drug delivery

Nanocrystalline and polycrystalline diamond films are also important for constructing biosensors. For example, NCD films patterned with O- and H-terminated microdomains were used for the construction of an impedance-based sensor for real-time monitoring of cell growth, and successfully applied in cultures of human osteoblast-like MG 63 cells [44]. An aptasensor for platelet-derived growth factor (PDGF) was designed on a NCD surface functionalized with carboxylic and amine groups, which served as units for immobilizing PDGF-binding aptamers [120]. Arrays based on polycrystalline diamond surfaces grafted with odorant binding proteins (i.e., small soluble proteins present in olfactory systems) were fabricated for potential artificial olfaction applications [121]. Electrodes coated with boron-doped diamond (BDD) films are also excellent materials for the construction of biosensors, especially after further modifications with other nanoparticles and molecules [122]. For example, a BDD electrode modified with silver nanoparticles was developed as a cholesterol sensor [123]. A biosensor for the detection of L-serine was fabricated using a BDD electrode modified with polycrystalline nickel and nickel(II) oxide (Ni-NiO) half nanotubes [124]. A BDD electrode modified with platinum nanoparticles (serving as a highly conductive catalytic transducer, and coupled with a competitive magneto-enzyme immunoassay), was used for the detection and degradation of the pesticide atrazine [125]. An electrochemical sensor for glutamate, an important neurotransmitter in the mammalian central nervous system, was fabricated on doped chemical vapor deposition diamond electrodes and graphene nanoplatelet structures [126]. Diamond coatings also enhanced the sensitivity and operation range of optical fiber sensors [127].

As for the drug delivery, oxidized ultrananocrystalline diamond thin films were functionalized with type I collagen and an anti-inflammatory drug dexamethasone, in order to fabricate a hybrid therapeutically active substrate for localized drug delivery [128].

4. Polymer-nanodiamond composites

Another promising group of materials for biomedical applications, particularly for bone tissue engineering, is polymer-nanodiamond composites. Polymers in general are too soft and elastic for bone tissue engineering, and thus they require reinforcements with harder and stronger materials, e.g., ceramic, metallic or carbon nanoparticles, including DNPs. Polymers reinforced with DNPs have been used in the form of thin films, e.g., poly(L-lactide) (PLLA) films [129, 130], or as electrospun nanofibrous scaffolds made of poly(lactide-co-glycolide) (PLGA) [43, 45]. Both types of composites showed improved mechanical properties and effectively supported the adhesion and growth of osteoblast-like MG 63 cells and human bone marrow MSCs (**Figure 5**). In the case of PLLA films, fluorescent DNPs were used in order to localize these materials in tissues after implantation and to investigate their behavior *in vivo* [129, 130]. In the case of nanofibrous PLGA scaffolds, the DNP-loaded scaffolds supported the growth of MG 63 cells to a similar degree as the pure scaffolds, but accelerated the growth of MSCs [43, 45]. This might be due to the fact that the primary or low-passaged cells, such as MSCs, were more sensitive to the physical and chemical properties of their growth substrate than the cell lines, which were better adapted to the *in vitro* conditions.

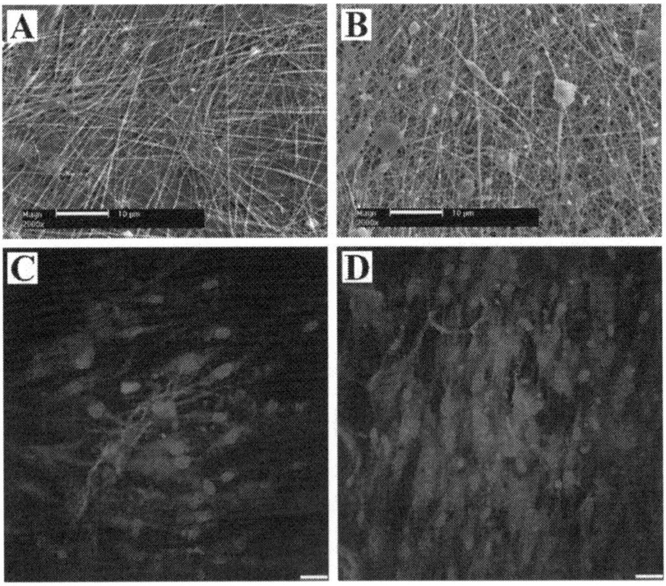

Figure 5. The morphology of pure PLGA nanofibrous scaffolds (A), PLGA scaffolds loaded with ~23 wt.% of diamond nanoparticles (B) and human bone marrow mesenchymal stem cells on day 9 after seeding on the PLGA (C) or PLGA-nanodiamond (D) scaffolds. A, B: scanning electron microscope, bar = 10 μm. C, D: cell membrane and cytoplasm stained with Texas Red C_2-Maleimide; cell nuclei counterstained with Hoechst #33342. Leica SPE confocal microscope, bar = 20 μm.

However, nanofibrous PLLA-nanodiamond scaffolds exerted rather adverse effects on their colonization with bone cells. The number, mitochondrial activity and expression of some markers of osteogenic differentiation at the mRNA and protein level in the human osteoblast-like MG 63 and Saos-2 cells cultured on these scaffolds decreased with the increasing concentration of DNPs in these scaffolds (from about 0.4 to 12 wt.%; [46]). At the same time, the concentration of DNPs in nanofibrous PLGA-nanodiamond scaffolds was considerably higher, i.e., ~23 wt.% [43, 45]. The different cell behavior on the PLGA- and PLLA-based scaffolds was probably due to the different origin and properties of the DNPs used for loading these scaffolds. For PLGA scaffolds, the DNPs were prepared using a radio frequency plasma activated chemical vapor deposition (RF PACVD) method [131], while for the PLLA scaffolds, pristine NanoAmando DNPs (see Section 2.4) were used. These DNPs were much smaller, hydrophobic, and induced cytotoxic effects in Saos-2 cells when added into the cell culture medium. They were probably released from the scaffolds, and may have also increased the hydrophobicity of the scaffolds, which could negatively affect cell functions.

DNPs have been also added in materials other than polymers. For example, these nanoparticles were used to increase the corrosion resistance of Mg, i.e., a material promising to construct biodegradable bone implants. Specifically, the corrosion rate of Mg was reduced by 4.5 times, when 5 wt.% of DNPs were uniformly dispersed in the Mg matrix. At the same time, L-929 fibroblasts cultured on the Mg-nanodiamond composites maintained high cell viability and a healthy morphology [132].

5. Conclusion

It can be concluded that diamond in the form of nanoparticles, nanostructured films and composite scaffolds, particularly nanofibrous scaffolds loaded with DNPs, is an exceptionally promising material for a wide range of biomedical applications. DNPs are suitable for drug and gene delivery, bioimaging, biosensing and also as additives to polymeric scaffolds for bone tissue engineering, particularly those nanofibrous, in order to improve their mechanical properties and increase their bioactivity. Nanodiamond films can be applied for localized drug delivery, construction of biosensors and particularly for bone implant coating due to their favorable effect on cell adhesion, growth and osteogenic differentiation. The cell performance on the nanodiamond films can be modulated by their roughness and topography, e.g., the size and shape of the surface irregularities, by termination with various atoms (e.g., O and H), chemical functional groups and biomolecules and by boron doping. However, free DNPs and DNPs dispersed in a material matrix should be applied with caution because of their higher reactivity with the surrounding environment and potential penetration into cells, which may finally result in their cytotoxicity.

Acknowledgements

Our studies included in this review were supported by the Agency for Healthcare Research, Ministry of Health of the Czech Republic (grant no. 15-32497A) and the Grant Agency of the Czech Republic (grant no. 14-04790S). Mrs. Paula Solon is gratefully acknowledged for her language revision of the manuscript.

Author details

Lucie Bacakova[1*], Antonin Broz[1], Jana Liskova[1], Lubica Stankova[1], Stepan Potocky[2] and Alexander Kromka[2]

*Address all correspondence to: lucie.bacakova@fgu.cas.cz

1 Institute of Physiology of the Czech Academy of Sciences, v.v.i., Prague, Czech Republic

2 Institute of Physics of the Czech Academy of Sciences, v.v.i., Prague, Czech Republic

References

[1] Badea I, Kaur R. Nanodiamonds as novel nanomaterials for biomedical applications: drug delivery and imaging systems. International Journal of Nanomedicine 2013; 8: 203. DOI: 10.2147/IJN.S37348.

[2] Shenderova OA, McGuire GE. Science and engineering of nanodiamond particle surfaces for biological applications (Review). Biointerphases 2015; 10(3): 030802. DOI: 10.1116/1.4927679.

[3] Field JE. The mechanical and strength properties of diamond. Reports on Progress in Physics 2012; 75(12): 126505. DOI: 10.1088/0034-4885/75/12/126505.

[4] Huang Q, Yu D, Xu B, Hu W, Ma Y, Wang Y, Zhao Z, Wen B, He J, Liu Z, Tian Y. Nanotwinned diamond with unprecedented hardness and stability. Nature 2014; 510(7504): 250–253. DOI: 10.1038/nature13381.

[5] Weber MJ. Handbook of optical materials. Boca Raton: CRC Press; 2002.

[6] Field JE, Pickles CSJ. Strength, fracture and friction properties of diamond. Diamond and Related Materials 1996; 5(6–8): 625–634. DOI: 10.1016/0925-9635(95)00362-2.

[7] Yang W, Auciello O, Butler JE, Cai W, Carlisle JA, Gerbi JE, Gruen DM, Knickerbocker T, Lasseter TL, Russell JN, Smith LM, Hamers RJ. DNA-modified nanocrystalline diamond thin-films as stable, biologically active substrates. Nature Materials 2002; 1(4): 253–257. DOI: 10.1038/nmat779.

[8] Härtl A, Schmich E, Garrido JA, Hernando J, Catharino SCR, Walter S, Feulner P, Kromka A, Steinmüller D, Stutzmann M. Protein-modified nanocrystalline diamond thin films for biosensor applications. Nature Materials 2004; 3(10): 736–742. DOI: 10.1038/nmat1204.

[9] Dubrovinskaia N, Dub S, Dubrovinsky L. Superior wear resistance of aggregated diamond nanorods. Nano Letters 2006; 6(4): 824–826. DOI: 10.1021/nl0602084.

[10] Silva FG, Neto MA, Fernandes AJS, Costa FM, Oliveira FJ, Silva RF. Adhesion and wear behaviour of NCD coatings on Si3N4 by micro-abrasion tests. Journal of Nanoscience and Nanotechnology 2009; 9(6): 3938–3943. DOI: 10.1166/jnn.2009.NS93.

[11] Ohfuji H, Irifune T, Litasov KD, Yamashita T, Isobe F, Afanasiev VP, Pokhilenko NP. Natural occurrence of pure nano-polycrystalline diamond from impact crater. Scientific Reports 2015; 5: 14702. DOI: 10.1038/srep14702.

[12] Shatskiy A, Yamazaki D, Morard G, Cooray T, Matsuzaki T, Higo Y, Funakoshi Ken-ichi, Sumiya H, Ito E, Katsura T. Boron-doped diamond heater and its application to large-volume, high-pressure, and high-temperature experiments. Review of Scientific Instruments 2009; 80(2): 023907. DOI: 10.1063/1.3084209.

[13] Booth L, Catledge SA, Nolen D, Thompson RG, Vohra YK. Synthesis and characterization of multilayered diamond coatings for biomedical implants. Materials 2011; 4(12): 857–868. DOI: 10.3390/ma4050857.

[14] Barletta M, Rubino G, Valle R, Polini R. Chemical vapor deposition of highly adherent diamond coatings onto co-cemented tungsten carbides irradiated by high power diode laser. ACS Applied Materials and Interfaces 2012; 4(2): 694–701. DOI: 10.1021/am2012992.

[15] Töyräs J, Korhonen RK, Voutilainen T, Jurvelin JS, Lappalainen R. Improvement of arthroscopic cartilage stiffness probe using amorphous diamond coating: application of amorphous diamond coating on an arthroscopic probe. Journal of Biomedical Materials Research, Part B: Applied Biomaterials 2005; 73B(1): 15–22. DOI: 10.1002/jbm.b.30173.

[16] Diamond Buying Guide 2016. http://guide.diamondpriceguru.com/diamond-and-ring-basics/4cs/how-are-colored-diamonds-formed/ [accessed January 20, 2016].

[17] Balasubramanian G, Lazariev A, Arumugam SR, Duan De-wen. Nitrogen-vacancy color center in diamond—emerging nanoscale applications in bioimaging and biosensing. Current Opinion in Chemical Biology 2014; 20: 69–77. DOI: 10.1016/j.cbpa.2014.04.014.

[18] Schirhagl R, Chang K, Loretz M, Degen CL. Nitrogen-vacancy centers in diamond: nanoscale sensors for physics and biology. Annual Review of Physical Chemistry 2014; 65(1): 83–105. DOI: 10.1146/annurev-physchem-040513-103659.

[19] Kratochvílová I, Ashcheulov P, Kovalenko A, Záliš S, Ledvina M, Mičová J. Luminescent diamond nanoparticles: physical, chemical and biological aspects of the phenomenon. Journal of Nanoscience and Nanotechnology 2015; 15(2): 1000–1005. DOI: 10.1166/jnn.2015.9740.

[20] Kromka A, Grausova L, Bacakova L, Vacik J, Rezek B, Vanecek M, Williams OA, Haenen K. Semiconducting to metallic-like boron doping of nanocrystalline diamond

films and its effect on osteoblastic cells. Diamond and Related Materials 2010; 19(2–3): 190–195. DOI: 10.1016/j.diamond.2009.10.003.

[21] Ekimov EA, Sidorov VA, Bauer ED, Mel'nik NN, Curro NJ, Thompson JD, Stishov SM. Superconductivity in diamond. Nature 2004; 428(6982): 542–545. DOI: 10.1038/nature02449.

[22] Kraft A. Doped diamond: a compact review on a new, versatile electrode material. International Journal of Electrochemical Science 2007; 2(5): 355–385.

[23] Bačáková L, Starý V, Kofronová O, Lisá V. Polishing and coating carbon fiber-reinforced carbon composites with a carbon-titanium layer enhances adhesion and growth of osteoblast-like MG63 cells and vascular smooth muscle cells in vitro. Journal of Biomedical Materials Research 2001; 54(4): 567–578. DOI: 10.1002/1097-4636(20010315)54:4<567::AID-JBM140>3.0.CO;2-Y.

[24] Steiner R, Beier US, Heiss-Kisielewsky I, Engelmeier R, Dumfahrt H, Dhima M. Adjusting dental ceramics: an in vitro evaluation of the ability of various ceramic polishing kits to mimic glazed dental ceramic surface. The Journal of Prosthetic Dentistry 2015; 113(6): 616–622. DOI: 10.1016/j.prosdent.2014.12.007.

[25] Jimbo H, Kamata S, Miura K, Masubuchi T, Ichikawa M, Ikeda Y, Haraoka J. En bloc temporal bone resection using a diamond threadwire saw for malignant tumors: technical note. Journal of Neurosurgery 2010: 1–4. DOI: 10.3171/2010.8.JNS10294.

[26] Wada K, Toyooka T, Ueno H, Tomura S, Tomiyama A, Mori K, Otani N, Fujii K. Safe fronto-orbito-zygomatic osteotomy using a diamond-coated threadwire saw in orbito-zygomatic craniotomy. Asian Journal of Neurosurgery 2015; 10(3): 250. DOI: 10.4103/1793-5482.161190.

[27] Jackson MJ, Sein H, Ahmed W, Woodwards R. Novel diamond-coated tools for dental drilling applications. Journal of Medical Engineering and Technology 2007; 31(2): 81–93. DOI: 10.1080/03091900500217489.

[28] Bacakova L, Grausova L, Vacik J, Fraczek A, Blazewicz S, Kromka A, Vanecek M, Svorcik V. Improved adhesion and growth of human osteoblast-like MG 63 cells on biomaterials modified with carbon nanoparticles. Diamond and Related Materials 2007; 16(12): 2133–2140. DOI: 10.1016/j.diamond.2007.07.015.

[29] Grausova L, Kromka A, Bacakova L, Potocky S, Vanecek M, Lisa V. Bone and vascular endothelial cells in cultures on nanocrystalline diamond films. Diamond and Related Materials 2008; 17(7–10): 1405–1409. DOI: 10.1016/j.diamond.2008.02.008.

[30] Grausova L, Bacakova L, Kromka A, Potocky S, Vanecek M, Nesladek M, Lisa V. Nanodiamond as promising material for bone tissue engineering. Journal of Nanoscience and Nanotechnology 2009; 9(6): 3524–3534. DOI: 10.1166/jnn.2009.NS26.

[31] Grausova L, Bacakova L, Kromka A, Vanecek M, Rezek B, Lisa V. Molecular markers of adhesion, maturation and immune activation of human osteoblast-like MG 63 cells

on nanocrystalline diamond films. Diamond and Related Materials 2009; 18(2–3): 258–263. DOI: 10.1016/j.diamond.2008.10.023.

[32] Grausova L, Kromka A, Burdikova Z, Eckhardt A, Rezek B, Vacik J, Haenen K, Lisa V, Bacakova L. Enhanced growth and osteogenic differentiation of human osteoblast-like cells on boron-doped nanocrystalline diamond thin films. PLoS One 2011; 6(6): e20943. DOI: 10.1371/journal.pone.0020943.

[33] Kopecek M, Bacakova L, Vacik J, Fendrych F, Vorlicek V, Kratochvilova I, Lisa V, Van Hove E, Mer C, Bergonzo P, Nesladek M. Improved adhesion, growth and maturation of human bone-derived cells on nanocrystalline diamond films. Physica Status Solidi (a) 2008; 205(9): 2146–2153. DOI: 10.1002/pssa.200879729.

[34] Broz A, Baresova V, Kromka A, Rezek B, Kalbacova M. Strong influence of hierarchically structured diamond nanotopography on adhesion of human osteoblasts and mesenchymal cells. Physica Status Solidi (a) 2009; 206(9): 2038–2041. DOI: 10.1002/pssa.200982203.

[35] Kalbacova M, Rezek B, Baresova V, Wolf-Brandstetter C, Kromka A. Nanoscale topography of nanocrystalline diamonds promotes differentiation of osteoblasts. Acta Biomaterialia 2009; 5(8): 3076–3085. DOI: 10.1016/j.actbio.2009.04.020.

[36] Kalbacova M, Broz A, Babchenko O, Kromka A. Study on cellular adhesion of human osteoblasts on nano-structured diamond films. Physica Status Solidi B 2009; 246(11–12): 2774–2777. DOI: 10.1002/pssb.200982253.

[37] Babchenko O, Kromka A, Hruska K, Kalbacova M, Broz A, Vanecek M. Fabrication of nano-structured diamond films for SAOS-2 cell cultivation. Physica Status Solidi (a) 2009; 206(9): 2033–2037. DOI: 10.1002/pssa.200982215.

[38] Rezek B, Michalíková L, Ukraintsev E, Kromka A, Kalbacova M. Micro-pattern guided adhesion of osteoblasts on diamond surfaces. Sensors 2009; 9(5): 3549–3562. DOI: 10.3390/s90503549.

[39] Rezek B, Ukraintsev E, Kromka A, Ledinský M, Brož A, Nosková L, Hartmannová H, Kalbacova M. Assembly of osteoblastic cell micro-arrays on diamond guided by protein pre-adsorption. Diamond and Related Materials 2010; 19(2–3): 153–157. DOI: 10.1016/j.diamond.2009.09.016.

[40] Ukraintsev E, Rezek B, Kromka A, Broz A, Kalbacova M. Long-term adsorption of fetal bovine serum on H/O-terminated diamond studied in situ by atomic force microscopy. Physica Status Solidi B 2009; 246(11–12): 2832–2835. DOI: 10.1002/pssb.200982257.

[41] Liskova J, Babchenko O, Varga M, Kromka A, Hadraba D, Svindrych Z, Burdikova Z, Bacakova L. Osteogenic cell differentiation on H-terminated and O-terminated nanocrystalline diamond films. International Journal of Nanomedicine 2015: 869. DOI: 10.2147/IJN.S73628.

[42] Liskova J, Babchenko O, Varga M, Kromka A, Hadraba D, Svindrych Z, Bacakova L. Effect of H- and O-termination of nanocrystalline diamond films on cell adhesion and

osteogenic differentiation. Nanocon 2014, 6th International Conference, November 5th-7th 2014, Brno, Czech Republic, EU, Conference Proceedings, 2015, TANGER Ltd., Ostrava, pages 650–656, ISBN 978-80-87294-53-6.

[43] Brady MA, Renzing A, Douglas TEL, Liu Q, Wille S, Parizek M, Bacakova L, Kromka A, Jarosova M, Godier G, Warnke PH. Development of composite poly(lactide-co-glycolide)-nanodiamond scaffolds for bone cell growth. Journal of Nanoscience and Nanotechnology 2015; 15(2): 1060–1069. DOI: 10.1166/jnn.2015.9745.

[44] Ižák T, Novotná K, Kopová I, Bačáková L, Rezek B, Kromka A. H-terminated diamond as optically transparent impedance sensor for real-time monitoring of cell growth: H-terminated diamond as optically transparent impedance sensor. Physica Status Solidi B 2013; 250(12): 2741–2746. DOI: 10.1002/pssb.201300098.

[45] Parizek M, Douglas TE, Novotna K, Kromka A, Brady MA, Renzing A, Voss E, Jarosova M, Palatinus L, Tesarek P, Ryparova P, Lisa V, dos Santos AM, Bacakova L. Nanofibrous poly(lactide-co-glycolide) membranes loaded with diamond nanoparticles as promising substrates for bone tissue engineering. International Journal of Nanomedicine 2012; 7: 1931. DOI: 10.2147/IJN.S26665.

[46] Bacakova L, Kopova I, Stankova L, Liskova J, Vacik J, Lavrentiev V, Kromka A, Potocky S, Stranska D. Bone cells in cultures on nanocarbon-based materials for potential bone tissue engineering: a review: bone cells in cultures on nanocarbon-based materials. Physica Status Solidi (a) 2014; 211(12): 2688–2702. DOI: 10.1002/pssa.201431402.

[47] Moosa B, Fhayli K, Li S, Julfakyan K, Ezzeddine A, Khashab NM. Applications of nanodiamonds in drug delivery and catalysis. Journal of Nanoscience and Nanotechnology 2014; 14(1): 332–343. DOI: 10.1166/jnn.2014.9141.

[48] Lee GJ, Kim CK, Bae Y, Rhee CK. Surface modification to improve hydrophobicity of detonation nanodiamond. Journal of Nanoscience and Nanotechnology 2012; 12(7): 5995–5999. DOI: 10.1166/jnn.2012.6345.

[49] Martín R, Álvaro M, Herance JR, García H. Fenton-treated functionalized diamond nanoparticles as gene delivery system. ACS Nano 2010; 4(1): 65–74. DOI: 10.1021/nn901616c.

[50] Desai C, Chen K, Mitra S. Aggregation behavior of nanodiamonds and their functionalized analogs in an aqueous environment. Environmental Science: Processes and Impacts 2014; 16(3): 518–523. DOI: 10.1039/C3EM00378G.

[51] Dhanak VR, Butenko YV, Brieva AC, Coxon PR, Alves L, Šiller L. Chemical functionalization of nanodiamond by amino groups: an x-ray photoelectron spectroscopy study. Journal of Nanoscience and Nanotechnology 2012; 12(4): 3084–3090. DOI: 10.1166/jnn.2012.4547.

[52] Badea I, Kaur R, Michel D, Chitanda JM, Maley J, Yang P, Borondics F, Verrall RE. Lysine-functionalized nanodiamonds: synthesis, physiochemical characterization, and

nucleic acid binding studies. International Journal of Nanomedicine 2012: 3851. DOI: 10.2147/IJN.S32877.

[53] Krueger A, Stegk J, Liang Y, Lu L, Jarre G. Biotinylated nanodiamond: simple and efficient functionalization of detonation diamond. Langmuir 2008; 24(8): 4200–4204. DOI: 10.1021/la703482v.

[54] Zhao L, Xu YH, Akasaka T, Abe S, Komatsu N, Watari F, Chen X. Polyglycerol-coated nanodiamond as a macrophage-evading platform for selective drug delivery in cancer cells. Biomaterials 2014; 35(20): 5393–5406. DOI: 10.1016/j.biomaterials.2014.03.041.

[55] Zhao L, Shiino A, Qin H, Kimura T, Komatsu N. Synthesis, characterization, and magnetic resonance evaluation of polyglycerol-functionalized detonation nanodiamond conjugated with gadolinium(III) complex. Journal of Nanoscience and Nanotechnology 2015; 15(2): 1076–1082. DOI: 10.1166/jnn.2015.9738.

[56] Cui C, Wang Y, Yang K, Wang Y, Yang J, Xi J, Zhao M, Wu J, Peng S. Preparation and characterization of RGDS/nanodiamond as a vector for VEGF-siRNA delivery. Journal of Biomedical Nanotechnology 2015; 11(1): 70–80. DOI: 10.1166/jbn.2015.2005.

[57] Ho D. Beyond the sparkle: the impact of nanodiamonds as biolabeling and therapeutic agents. ACS Nano 2009; 3(12): 3825–3829. DOI: 10.1021/nn9016247.

[58] Faklaris O, Joshi V, Irinopoulou T, Tauc P, Sennour M, Girard H, Gesset C, Arnault JC, Thorel A, Boudou JP, Curmi PA, Treussart F. Photoluminescent diamond nanoparticles for cell labeling: study of the uptake mechanism in mammalian cells. ACS Nano 2009; 3(12): 3955–3962. DOI: 10.1021/nn901014j.

[59] Bertrand JR, Pioche-Durieu C, Ayala J, Petit T, Girard HA, Malvy CP, Le Cam E, Treussart F, Arnault JC. Plasma hydrogenated cationic detonation nanodiamonds efficiently deliver to human cells in culture functional siRNA targeting the Ewing sarcoma junction oncogene. Biomaterials 2015; 45: 93–98. DOI: 10.1016/j.biomaterials.2014.12.007.

[60] Cheng LC, Chen HM, Lai TC, Chan YC, Liu RS, Sung JC, Hsiao M, Chen CH, Her LJ, Tsai DP. Targeting polymeric fluorescent nanodiamond-gold/silver multi-functional nanoparticles as a light-transforming hyperthermia reagent for cancer cells. Nanoscale 2013; 5(9): 3931. DOI: 10.1039/c3nr34091k.

[61] Hsu MH, Chuang H, Cheng FY, Huang YP, Han CC, Chen JY, Huang SC, Chen JK, Wu DS, Chu HL, Chang CC. Directly thiolated modification onto the surface of detonation nanodiamonds. ACS Applied Materials and Interfaces 2014; 6(10): 7198–7203. DOI: 10.1021/am500324z.

[62] Man HB, Kim H, Kim HJ, Robinson E, Liu WK, Chow EKH, Ho D. Synthesis of nanodiamond–daunorubicin conjugates to overcome multidrug chemoresistance in leukemia. Nanomedicine: Nanotechnology, Biology and Medicine 2014; 10(2): 359–369. DOI: 10.1016/j.nano.2013.07.014.

[63] Li J, Zhu Y, Li W, Zhang X, Peng Y, Huang Q. Nanodiamonds as intracellular transporters of chemotherapeutic drug. Biomaterials 2010; 31(32): 8410–8418. DOI: 10.1016/j.biomaterials.2010.07.058.

[64] Mochalin VN, Pentecost A, Li XM, Neitzel I, Nelson M, Wei C, He T, Guo F, Gogotsi Y. Adsorption of drugs on nanodiamond: toward development of a drug delivery platform. Molecular Pharmaceutics 2013; 10(10): 3728–3735. DOI: 10.1021/mp400213z.

[65] Huang H, Pierstorff E, Osawa E, Ho D. Protein-mediated assembly of nanodiamond hydrogels into a biocompatible and biofunctional multilayer nanofilm. ACS Nano 2008; 2(2): 203–212. DOI: 10.1021/nn7000867.

[66] Vaijayanthimala V, Lee DK, Kim SV, Yen A, Tsai N, Ho D, Chang HC, Shenderova O. Nanodiamond-mediated drug delivery and imaging: challenges and opportunities. Expert Opinion on Drug Delivery 2015; 12(5): 735–749. DOI: 10.1517/17425247.2015.992412.

[67] Xiao J, Duan X, Yin Q, Zhang Z, Yu H, Li Y. Nanodiamonds-mediated doxorubicin nuclear delivery to inhibit lung metastasis of breast cancer. Biomaterials 2013; 34(37): 9648–9656. DOI: 10.1016/j.biomaterials.2013.08.056.

[68] Xi G, Robinson E, Mania-Farnell B, Vanin EF, Shim KW, Takao T, Allender EV, Mayanil CS, Soares MB, Ho D, Tomita T. Convection-enhanced delivery of nanodiamond drug delivery platforms for intracranial tumor treatment. Nanomedicine: Nanotechnology, Biology and Medicine 2014; 10(2): 381–391. DOI: 10.1016/j.nano.2013.07.013.

[69] Moore L, Gatica M, Kim H, Osawa E, Ho D. Multi-protein delivery by nanodiamonds promotes bone formation. Journal of Dental Research 2013; 92(11): 976–981. DOI: 10.1177/0022034513504952.

[70] Chu HL, Chen HW, Tseng SH, Hsu MH, Ho LP, Chou FH, Li MPHY, Chang YC, Chen PH, Tsai LY, Chou CC, Chen JS, Cheng TM, Chang CC. Development of a growth-hormone-conjugated nanodiamond complex for cancer therapy. ChemMedChem 2014; 9(5): 1023–1029. DOI: 10.1002/cmdc.201300541.

[71] Smith AH, Robinson EM, Zhang XQ, Chow EK, Lin Y, Osawa E, Xi J, Ho D. Triggered release of therapeutic antibodies from nanodiamond complexes. Nanoscale 2011; 3(7): 2844–2848. DOI: 10.1039/c1nr10278h.

[72] Yan L, Yang Y, Zhang W, Chen X. Advanced materials and nanotechnology for drug delivery. Advanced Materials 2014; 26(31): 5533–5540. DOI: 10.1002/adma.201305683.

[73] Eidi H, David MO, Crépeaux G, Henry L, Joshi V, Berger MH, Sennour M, Cadusseau J, Gherardi RK, Curmi PA. Fluorescent nanodiamonds as a relevant tag for the assessment of alum adjuvant particle biodisposition. BMC Medicine 2015; 13(1). DOI: 10.1186/s12916-015-0388-2.

[74] Lam R, Ho D. Nanodiamonds as vehicles for systemic and localized drug delivery. Expert Opinion on Drug Delivery 2009; 6(9): 883–895. DOI: 10.1517/17425240903156382.

[75] Wei L, Zhang W, Lu H, Yang P. Immobilization of enzyme on detonation nanodiamond for highly efficient proteolysis. Talanta 2010; 80(3): 1298–1304. DOI: 10.1016/j.talanta.2009.09.029.

[76] Wang Z, Yang Y, Xu Z, Wang Y, Zhang W, Shi P. Interrogation of cellular innate immunity by diamond-nanoneedle-assisted intracellular molecular fishing. Nano Letters 2015; 15(10): 7058–7063. DOI: 10.1021/acs.nanolett.5b03126.

[77] Wu TJ, Tzeng YK, Chang WW, Cheng CA, Kuo Y, Chien CH, Chang HC, Yu J. Tracking the engraftment and regenerative capabilities of transplanted lung stem cells using fluorescent nanodiamonds. Nature Nanotechnology 2013; 8(9): 682–689. DOI: 10.1038/nnano.2013.147.

[78] Khalid A, Chung K, Rajasekharan R, Lau DWM, Karle TJ, Gibson BC, Tomljenovic-Hanic S. Lifetime reduction and enhanced emission of single photon color centers in nanodiamond via surrounding refractive index modification. Scientific Reports 2015; 5: 11179. DOI: 10.1038/srep11179.

[79] Zhang T, Cui H, Fang CY, Su LJ, Ren S, Chang HC, Yang X, Forrest ML. Photoacoustic contrast imaging of biological tissues with nanodiamonds fabricated for high near-infrared absorbance. Journal of Biomedical Optics 2013; 18(2): 026018. DOI: 10.1117/1.JBO.18.2.026018.

[80] Prabhakar N, Näreoja T, von Haartman E, Karaman DŞ, Jiang H, Koho S, Dolenko TA, Hänninen PE, Vlasov DI, Ralchenko VG, Hosomi S, Vlasov II, Sahlgren C, Rosenholm JM. Core–shell designs of photoluminescent nanodiamonds with porous silica coatings for bioimaging and drug delivery II: application. Nanoscale 2013; 5(9): 3713. DOI: 10.1039/c3nr33926b.

[81] Briones M, Casero E, Petit-Domínguez MD, Ruiz MA, Parra-Alfambra AM, Pariente F, Lorenzo E, Vázquez L. Diamond nanoparticles based biosensors for efficient glucose and lactate determination. Biosensors and Bioelectronics 2015; 68: 521–528. DOI: 10.1016/j.bios.2015.01.044.

[82] Bacakova L, Grausova L, Vandrovcova M, Vacik J, Frazcek A, Blazewicz S, Kromka, A, Vanecek M, Nesladek M, Svorcik V, Kopecek M: Carbon nanoparticles as substrates for cell adhesion and growth. In: Nanoparticles: New Research (Ed. Simone Luca Lombardi), Nova Science Publishers, Inc., Hauppauge, New York, USA; 2008, 39–107, ISBN 978-1-60456-704-5

[83] Bacakova L, Grausova L, Vacik J, Kromka A, Biederman H, Choukourov A, Stary V. Nanocomposite and nanostructured carbon-based films as growth substrates for bone cells. In: Reddy B, editor. Advances in Diverse Industrial Applications of Nanocomposites, InTech; 2011, pp 399–436, ISBN 978-953-307-202-9.

[84] Blaber SP, Hill CJ, Webster RA, Say JM, Brown LJ, Wang SC, Vesey G, Herbert BR. Effect of labeling with iron oxide particles or nanodiamonds on the functionality of

adipose-derived mesenchymal stem cells. PLoS One 2013; 8(1): e52997. DOI: 10.1371/journal.pone.0052997.

[85] Puzyr AP, Tarskikh SV, Makarskaya GV, Chiganova GA, Larionova IS, Detkov PY, Bondar VS. Damaging effect of detonation diamonds on human white and red blood cells in vitro. Doklady Biochemistry and Biophysics 2002; 385: 201–204. DOI: 10.1023/A:1019959322589.

[86] Puzyr AP, Neshumaev DA, Tarskikh SV, Makarskaia GV, Dolmatov VI, Bondar' VS. Destruction of human blood cells upon interaction with detonation nanodiamonds in experiments in vitro. Biofizika 2005; 50(1): 101–106.

[87] Karpukhin AV, Avkhacheva NV, Yakovlev RY, Kulakova II, Yashin VA, Lisichkin GV, Safronova VG. Effect of detonation nanodiamonds on phagocyte activity. Cell Biology International 2011; 35(7): 727–733. DOI: 10.1042/CBI20100548.

[88] Niemiec T, Szmidt M, Sawosz E, Grodzik M, Mitura K. The effect of diamond nanoparticles on redox and immune parameters in rats. Journal of Nanoscience and Nanotechnology 2011; 11(10): 9072–9077. DOI: 10.1166/jnn.2011.3511.

[89] Kumari S, Singh MK, Singh SK, Grácio JJ, Dash D. Nanodiamonds activate blood platelets and induce thromboembolism. Nanomedicine 2014; 9(3): 427–440. DOI: 10.2217/nnm.13.23.

[90] Chwalibog A, Wierzbicki M, Sawosz E, Grodzik M, Hotowy A, Prasek M, Jaworski S, Sawosz F. Carbon nanoparticles downregulate expression of basic fibroblast growth factor in the heart during embryogenesis. International Journal of Nanomedicine 2013: 3427. DOI: 10.2147/IJN.S49745.

[91] Solarska K, Gajewska A, Bartosz G, Mitura K. Induction of apoptosis in human endothelial cells by nanodiamond particles. Journal of Nanoscience and Nanotechnology 2012; 12(6): 5117–5121. DOI: 10.1166/jnn.2012.4952.

[92] Solarska-Ściuk K, Gajewska A, Skolimowski J, Gajek A, Bartosz G. Cellular redox homeostasis in endothelial cells treated with nonmodified and Fenton-modified nanodiamond powders: cellular redox homeostasis in endothelial cells. Biotechnology and Applied Biochemistry 2014; 61(5): 593–602. DOI: 10.1002/bab.1208.

[93] Xing Y, Xiong W, Zhu L, Osawa E, Hussin S, Dai L. DNA damage in embryonic stem cells caused by nanodiamonds. ACS Nano 2011; 5(3): 2376–2384. DOI: 10.1021/nn200279k.

[94] Chwalibog A, Hinzmann M, Kutwin M, Jagiello J, Kozinski R, Wierzbicki M, Grodzik M, Lipinska L, Sawosz E, Jaworski S. Nanoparticles containing allotropes of carbon have genotoxic effects on glioblastoma multiforme cells. International Journal of Nanomedicine 2014: 2409. DOI: 10.2147/IJN.S62497.

[95] Mytych J, Lewinska A, Bielak-Zmijewska A, Grabowska W, Zebrowski J, Wnuk M. Nanodiamond-mediated impairment of nucleolar activity is accompanied by oxidative

stress and DNMT2 upregulation in human cervical carcinoma cells. Chemico-Biological Interactions 2014; 220: 51–63. DOI: 10.1016/j.cbi.2014.06.004.

[96] Keremidarska M, Ganeva A, Mitev D, Hikov T, Presker R, Pramatarova L, Krasteva N. Comparative study of cytotoxicity of detonation nanodiamond particles with an osteosarcoma cell line and primary mesenchymal stem cells. Biotechnology and Biotechnological Equipment 2014; 28(4): 733–739. DOI: 10.1080/13102818.2014.947704.

[97] Zhu Y, Li W, Zhang Y, Li J, Liang L, Zhang X, Chen N, Sun Y, Chen W, Tai R, Fan C, Huang Q. Excessive sodium ions delivered into cells by nanodiamonds: implications for tumor therapy. Small 2012; 8(11): 1771–1779. DOI: 10.1002/smll.201102539.

[98] Solarska-Ściuk K, Gajewska A, Skolimowski J, Mitura K, Bartosz G. Stimulation of production of reactive oxygen and nitrogen species in endothelial cells by unmodified and Fenton-modified ultradisperse detonation diamond: effect of nanopowders on the redox homeostasis in cells. Biotechnology and Applied Biochemistry 2013; 60(2): 259–265. DOI: 10.1002/bab.1071.

[99] Xing Z, Pedersen TO, Wu X, Xue Y, Sun Y, Finne-Wistrand A, Kloss FR, Waag T, Krueger A, Steinmüller-Nethl D, Mustafa K. Biological effects of functionalizing copolymer scaffolds with nanodiamond particles. Tissue Engineering, Part A 2013; 19(15–16): 1783–1791. DOI: 10.1089/ten.tea.2012.0336.

[100] Kromka A, Rezek B, Remes Z, Michalka M, Ledinsky M, Zemek J, Potmesil J, Vanecek M. Formation of continuous nanocrystalline diamond layers on glass and silicon at low temperatures. Chemical Vapor Deposition 2008; 14(7–8): 181–186. DOI: 10.1002/cvde.200706662.

[101] Prokhorenkov VI, Vasil'eva EY, Puzyr' AP, Bondar' VS. Effects of nanodiamonds of explosive synthesis on the skin of experimental animals locally exposed to cobalt and chrome ions. Bulletin of Experimental Biology and Medicine 2014; 158(2): 264–267. DOI: 10.1007/s10517-014-2737-0.

[102] Hopper AP, Dugan JM, Gill AA, Fox OJL, May PW, Haycock JW, Claeyssens F. Amine functionalized nanodiamond promotes cellular adhesion, proliferation and neurite outgrowth. Biomedical Materials 2014; 9(4): 045009. DOI: 10.1088/1748-6041/9/4/045009.

[103] Chen YC, Lee DC, Hsiao CY, Chung YF, Chen HC, Thomas JP, Pong WF, Tai NH, Lin IN, Chiu IM. The effect of ultra-nanocrystalline diamond films on the proliferation and differentiation of neural stem cells. Biomaterials 2009; 30(20): 3428–3435. DOI: 10.1016/j.biomaterials.2009.03.058.

[104] Chen YC, Lee DC, Tsai TY, Hsiao CY, Liu JW, Kao CY, Lin HK, Chen HC, Palathinkal TJ, Pong WF, Tai NH, Lin IN, Chiu IM. Induction and regulation of differentiation in neural stem cells on ultra-nanocrystalline diamond films. Biomaterials 2010; 31(21): 5575–5587. DOI: 10.1016/j.biomaterials.2010.03.061.

[105] Lechleitner T, Klauser F, Seppi T, Lechner J, Jennings P, Perco P, Mayer B, Steinmüller-Nethl D, Preiner J, Hinterdorfer P, Hermann M, Bertel E, Pfaller K, Pfaller W. The surface properties of nanocrystalline diamond and nanoparticulate diamond powder and their suitability as cell growth support surfaces. Biomaterials 2008; 29(32): 4275–4284. DOI: 10.1016/j.biomaterials.2008.07.023.

[106] Amaral M, Gomes PS, Lopes MA, Santos JD, Silva RF, Fernandes MH. Cytotoxicity evaluation of nanocrystalline diamond coatings by fibroblast cell cultures. Acta Biomaterialia 2009; 5(2): 755–763. DOI: 10.1016/j.actbio.2008.08.015.

[107] Bacakova L, Filova E, Liskova J, Kopova I, Vandrovcova M, Havlikova J. Nanostructured materials as substrates for the adhesion, growth, and osteogenic differentiation of bone cells. In: Nanobiomaterials in Hard Tissue Engineering. Applications of Nanobiomaterials. Volume 4, Ed. A. M. Grumezescu, Elsevier Inc., 2016, William Andrew Publishing, Oxford, Cambridge; Chapter 4, pp. 103–153, ISBN 978-0-323-42862-0

[108] Amaral M, Dias AG, Gomes PS, Lopes MA, Silva RF, Santos JD, Fernandes MH. Nanocrystalline diamond: in vitro biocompatibility assessment by MG63 and human bone marrow cells cultures. Journal of Biomedical Materials Research, Part A 2008; 87A(1): 91–99. DOI: 10.1002/jbm.a.31742.

[109] Clem WC, Chowdhury S, Catledge SA, Weimer JJ, Shaikh FM, Hennessy KM, Konovalov VV, Hill MR, Waterfeld A, Bellis SL, Vohra YK. Mesenchymal stem cell interaction with ultra-smooth nanostructured diamond for wear-resistant orthopaedic implants. Biomaterials 2008; 29(24–25): 3461–3468. DOI: 10.1016/j.biomaterials.2008.04.045.

[110] Webster TJ, Ergun C, Doremus RH, Siegel RW, Bizios R. Specific proteins mediate enhanced osteoblast adhesion on nanophase ceramics. Journal of Biomedical Materials Research 2000; 51(3): 475–483. DOI: 10.1002/1097-4636(20000905)51:3<475::AID-JBM23>3.0.CO;2–9.

[111] Webster TJ, Schadler LS, Siegel RW, Bizios R. Mechanisms of enhanced osteoblast adhesion on nanophase alumina involve vitronectin. Tissue Engineering 2001; 7(3): 291–301. DOI: 10.1089/10763270152044152.

[112] Bacakova L, Filova E, Parizek M, Ruml T, Svorcik V. Modulation of cell adhesion, proliferation and differentiation on materials designed for body implants. Biotechnology Advances 2011; 29(6): 739–767. DOI: 10.1016/j.biotechadv.2011.06.004.

[113] Yang L, Sheldon BW, Webster TJ. Orthopedic nano diamond coatings: control of surface properties and their impact on osteoblast adhesion and proliferation. Journal of Biomedical Materials Research, Part A 2009; 91A(2): 548–556. DOI: 10.1002/jbm.a.32227.

[114] Yang L, Chinthapenta V, Li Q, Stout D, Liang A, Sheldon BW, Webster TJ. Understanding osteoblast responses to stiff nanotopographies through experiments and

computational simulations. Journal of Biomedical Materials Research, Part A 2011; 97A(4): 375–382. DOI: 10.1002/jbm.a.33094.

[115] Pareta R, Yang L, Kothari A, Sirinrath S, Xiao X, Sheldon BW, Webster TJ. Tailoring nanocrystalline diamond coated on titanium for osteoblast adhesion. Journal of Biomedical Materials Research, Part A 2010; 95A(1): 129–136. DOI: 10.1002/jbm.a.32821.

[116] Hill MR, Catledge SA, Konovalov V, Clem WC, Chowdhury SA, Etheridge BS, Stanishevsky A, Lemons JE, Vohra YK, Eberhardt AW. Preliminary tribological evaluation of nanostructured diamond coatings against ultra-high molecular weight polyethylene. Journal of Biomedical Materials Research, Part B: Applied Biomaterials 2008; 85B(1): 140–148. DOI: 10.1002/jbm.b.30926.

[117] Duailibi SE, Duailibi MT, Ferreira LM, Salmazi KI, Salvadori MC, de Sa Teixeira F, Pasquarelli A, Vacanti JP, Yelick PC. Tooth tissue engineering: the influence of hydrophilic surface on nanocrystalline diamond films for human dental stem cells. Tissue Engineering, Part A 2013; 19(23–24): 2537–2543. DOI: 10.1089/ten.tea.2012.0628.

[118] Yang L, Li Y, Sheldon BW, Webster TJ. Altering surface energy of nanocrystalline diamond to control osteoblast responses. Journal of Materials Chemistry 2012; 22(1): 205–214. DOI: 10.1039/C1JM13593G.

[119] Hristova K, Pecheva E, Pramatarova L, Altankov G. Improved interaction of osteoblast-like cells with apatite–nanodiamond coatings depends on fibronectin. Journal of Materials Science: Materials in Medicine 2011; 22(8): 1891–1900. DOI: 10.1007/s10856-011-4357-9.

[120] Wang X, Ishii Y, Ruslinda AR, Hasegawa M, Kawarada H. Effective surface functionalization of nanocrystalline diamond films by direct carboxylation for PDGF detection via aptasensor. ACS Applied Materials and Interfaces 2012; 4(7): 3526–3534. DOI: 10.1021/am300595f.

[121] Manai R, Scorsone E, Rousseau L, Ghassemi F, Possas Abreu M, Lissorgues G, Tremillon N, Ginisty H, Arnault JC, Tuccori E, Bernabei M, Cali K, Persaud KC, Bergonzo P. Grafting odorant binding proteins on diamond bio-MEMS. Biosensors and Bioelectronics 2014; 60: 311–317. DOI: 10.1016/j.bios.2014.04.020.

[122] Svítková J, Ignat T, Švorc Ľ, Labuda J, Barek J. Chemical modification of boron-doped diamond electrodes for applications to biosensors and biosensing. Critical Reviews in Analytical Chemistry 2016; 46(3): 248–56. DOI: 10.1080/10408347.2015.1082125.

[123] Nantaphol S, Chailapakul O, Siangproh W. A novel paper-based device coupled with a silver nanoparticle-modified boron-doped diamond electrode for cholesterol detection. Analytica Chimica Acta 2015; 891: 136–143. DOI: 10.1016/j.aca.2015.08.007.

[124] Dai W, Li H, Li M, Li C, Wu X, Yang B. Electrochemical imprinted polycrystalline nickel–nickel oxide half-nanotube-modified boron-doped diamond electrode for the

detection of L-serine. ACS Applied Materials and Interfaces 2015; 7(41): 22858–22867. DOI: 10.1021/acsami.5b05642.

[125] Medina-Sánchez M, Mayorga-Martinez CC, Watanabe T, Ivandini TA, Honda Y, Pino F, Nakata K, Fujishima A, Einaga Y, Merkoçi A. Microfluidic platform for environmental contaminants sensing and degradation based on boron-doped diamond electrodes. Biosensors and Bioelectronics 2016; 75: 365–374. DOI: 10.1016/j.bios.2015.08.058.

[126] Hu J, Wisetsuwannaphum S, Foord JS. Glutamate biosensors based on diamond and graphene platforms. Faraday Discussions 2014; 172:457–72. DOI: 10.1039/C4FD00032C.

[127] Neto VF, Santos JA, Alberto NJ, Pinto JL, Nogueira RN, Grácio J. Evaluation of diamond coatings on optical fibre sensors for biological use. Journal of Nanoscience and Nanotechnology 2011; 11(6): 5408–5412. DOI: 10.1166/jnn.2011.3774.

[128] Huang H, Chen M, Bruno P, Lam R, Robinson E, Gruen D, Ho D. Ultrananocrystalline diamond thin films functionalized with therapeutically active collagen networks. The Journal of Physical Chemistry, B 2009; 113(10): 2966–2971. DOI: 10.1021/jp9004086.

[129] Zhang Q, Mochalin VN, Neitzel I, Knoke IY, Han J, Klug CA, Zhou JG, Lelkes PI, Gogotsi Y. Fluorescent PLLA-nanodiamond composites for bone tissue engineering. Biomaterials 2011; 32(1): 87–94. DOI: 10.1016/j.biomaterials.2010.08.090.

[130] Zhang Q, Mochalin VN, Neitzel I, Hazeli K, Niu J, Kontsos A, Zhou JG, Lelkes PI, Gogotsi Y. Mechanical properties and biomineralization of multifunctional nanodiamond-PLLA composites for bone tissue engineering. Biomaterials 2012; 33(20): 5067–5075. DOI: 10.1016/j.biomaterials.2012.03.063.

[131] Mitura S, Mitura K, Niedzielski P, Louda P, Danilenko V. Nanocrystalline diamond, its synthesis, properties and applications. Journal of Achievements in Materials and Manufacturing Engineering 2006; 16(1–2): 9–16.

[132] Gong H, Anasori B, Dennison CR, Wang K, Kumbur EC, Strich R, Zhou JG. Fabrication, biodegradation behavior and cytotoxicity of Mg-nanodiamond composites for implant application. Journal of Materials Science: Materials in Medicine 2015; 26(2). DOI: 10.1007/s10856-015-5441-3.

Chapter 4

Laser- Assisted Machining of Difficult to Cut Materials

Damian Przestacki

Additional information is available at the end of the chapter

http://dx.doi.org/10.5772/63417

Abstract

The chapter presents the results of experimental investigation of the machinability of difficult to cut materials, e.g. silicon carbide particulate, aluminium metal matrix composite, cemented carbides layers, technical ceramics (Si_3N_4). The work focuses on the study of machinability during conventional and laser-assisted turning. Nowadays, the difficult to cut materials have a great application in different fields of industry, e.g. automobile or aerospace. Metal matrix composites are known as the difficult to machine materials, because of the hardness and abrasive nature element—in this case —SiC.

The work reported concentrates on the machinability improvement of difficult to cut materials by laser-assisted machining (LAM) when compared with conventional cutting process. Influence of laser beam during laser-assisted turning on temperature of the heating area, tool wear, and machined surfaces' roughness was investigated. This research was carried out for polycrystalline diamond wedges (PCD), ceramics and sintered carbide wedges with coatings and uncoated ones. The results obtained with the laser assisted machining were compared with results obtained in conventional turning.

Keywords: laser-assisted machining, polycrystalline diamond, technical ceramics, metal matrix composites, cemented carbides

1. Introduction

The metal matrix composites as well as Si_3N_4 ceramics are important materials with many engineering and medical applications. However, the high hardness (2125HV) and brittleness of ceramics and the abrasive characteristics of the reinforced particulates in metal matrix composites (MMC) make serious difficulties when machining [1–3].

Mechanical and physical properties like high thermal resistance, high hardness, good corrosive resistance and chemical stability have encouraged use of ceramics and SiC reinforced aluminium-based composites in several engineering applications. The brittle and hard nature of these ceramics makes them difficult to machine using conventional techniques and damage caused to the surface while machining affects efficiency of components. Laser-assisted machining (LAM) has recently emerged as a potential technique for attaining high material removal rates.

MMCs and silicon nitride are increasingly attractive for engineering applications because of their superior combination of properties such as high strength, hardness, good corrosive resistance, and wear resistance. These materials are finding potential applications in the manufacture of automotive components such as some elements of cars, aerospace and medical [4, 5]. However, because of the high hardness of ceramic reinforcement, conventional turning and diamond machining, which represent the most widely used machining methods, currently provide the only options for machining metal matrix composites to the required accuracy. Even for the cutting wedges made of polycrystalline diamond and cubic boron nitride the tool wear is intense [3, 6]. Nevertheless, this technique involves low material removal rates and high tool wear, which result in high machining cost. Therefore, there is still a demand for hard to machine materials processing methods capable of enhancing high material removal rates, improving tool wear, and increasing the surface quality of the workpiece [7–9].

One way to improve the machinability of difficult to cut materials is to employ the thermal softening ability of a heat source to fluxing the material. The heat source is focused in front of a cutting tool to soften the workpiece material just ahead of the cutting tool thereby lowering the forces required to cut the material [1]. The technology is called a hybrid process of the machining, where the material is heated via laser irradiation and soft surface layer is machined by defined cutting edge of the wedges. This makes the material softer and easier to plastic deformation. In comparison with conventional machining this method significantly increases wedges durability [1, 7, 10–13]. LAM reduces the tool wear and cost of machining by reducing machine hours per part [14, 15]. In a study, Chang and Kuo [10] found a reduction of 20–22% in cutting force, which is an indicator of reducing the tool wear, and an increase in the profitability of using LAM during planning of alumina ceramics.

Research presented in this chapter builds on previous work by the author to further understand the process of laser-assisted turning. The present paper describes the experimental setup and characterization of a LAM process of a Si_3N_4 ceramics, cemented carbide layer, and metal matrix composites with a particular focus on the diamond tool wear and surface roughness.

2. Experiments

In the research, the standard metal matrix composite produced by the DURALCAN company was used as the workpiece material. The composite of a chemical composition similar to the AlSi9Mg alloy (9.2% Si, 0.6% Mg, 0.1% Fe) is reinforced with SiC particles with a particle size of between 8 to 15 μm and mass fraction about 20% (**Figure 1**). The sample has 50 mm in diameter and 10 mm length and before LAM it was painted by absorptive coat to improve absorbability of laser beam.

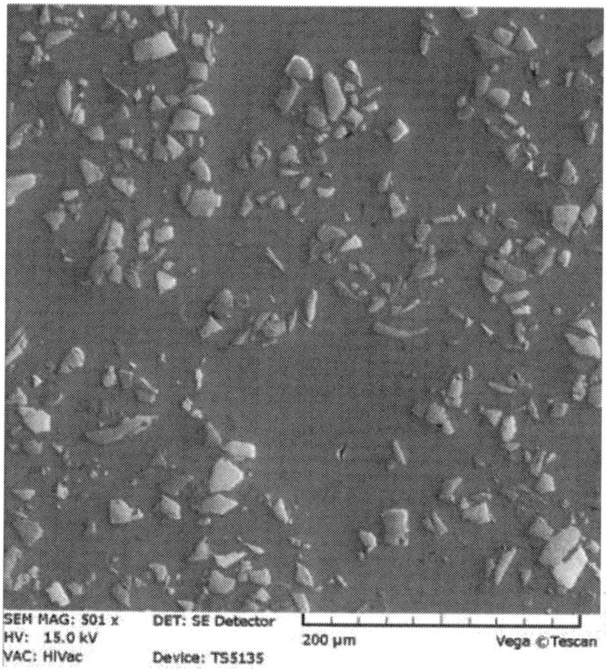

Figure 1. Microstructure of metal matrix composite used in research. The SiC particles are distributed through the matrix.

The machining tests were carried out on TUM 35D1 lathe with infinitely variable adjustment of rotational speed.

In order to determine the laser heating conditions for materials, there were materials that have been investigated, presented in **Table 1**.

Kind of masterials	Hardness	Grain size (μm)
AlSi9Mg (9.2% Si, 0.6% Mg, 0.1% Fe) +20% SiC	77 HRB	8–12
Cemented carbide (60% WC, 40% NiCr)	3000±500 HV	40–100
Si_3N_4	1809±40 $HV_{0,5}$	2–4

Table 1. Characteristics of the samples.

The example of laser-assisted turning of ceramics Si_3N_4 is illustrated in **Figure 2**.

The experimental setup applied for turning of Si_3N_4 ceramics is presented in **Figure 3**, while the setup for turning of MMC in LAM conditions is presented in **Figure 4**.

Figure 2. View of ceramics Si_3N_4 (Ø14×4.5×104mm) heated by laser beam.

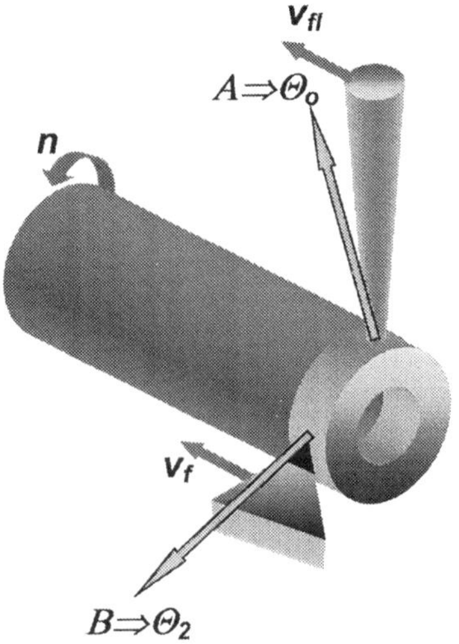

Figure 3. The scheme of experimental set-up. Ceramics turning with LAM, designations: *A*—heating area by laser beam, *B*—zone of machining, *d*—workpiece diameter [1].

http://dx.doi.org/10.5772/63417

The 30 degree angle between heated area and zone of machining was selected **Figure 4**. The tools wear were measured for each trial of experiment. The surface roughness of the sample was measured by surface analyzer (Hommel T500). The machined surface finish was characterized by the average arithmetic roughness (R*a* parameter). Tool wear was measured on the primary flank face by the optical microscope.

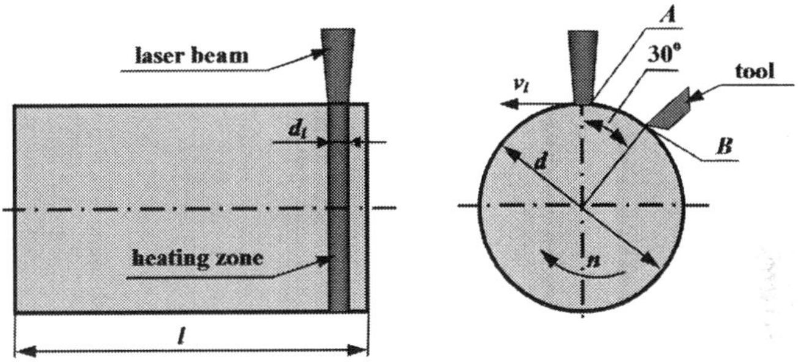

Figure 4. The scheme of experimental setup. (a) MMC laser-assisted turning, (b) Ceramics turning with LAM, designations: *A*—heating area by laser beam, *B*—zone of machining, *d*—workpiece diameter, d_l—laser beam diameter [1].

LAM was carried out with a 2.6 kW CO_2 laser (2.6kW Trumpf, type TLF2600t).

The workstation for Si_3N_4 ceramics was equipped with remotely controlled temperature measurement system, which works on the base of infrared detection of irradiation. The measurement data logging gives an opportunity of on-line tracking temperatures in: θ_0—machined workpiece temperature in laser beam power spot size on the heated surface; θ_2—cutting tool point on the machined surface temperature **Figure 3**). The laser-assisted turning process was done in two steps. The ceramic bush with n rotational speed was heated in t_o period time up to reach the target of suitable workpiece temperature. The next step was the initiation of feed motion, and then the laser power spot size within wedge tool was straight moved. The constant temperature during laser-assisted turning process was ensured by dedicated computer program. The obtained θ_2 temperature data logging was compared with manual typing process temperature. The on-line program has changed laser beam power in case of appeared differences between those temperatures. The system power control was worked on the base of negative feedback.

The research was carried out for these cutting parameters:

Si_3N_4: cutting speed v_c = 10 m/min, feed rate f = 0.04mm/rev, depth of cut a_p = 0.05mm;

MMC:cutting speed v_c = 10–100m/min, feed rate f = 0.04–0.1 mm/rev, depth of cut a_p = 0.1mm;

Cemented carbide: cutting speed v_c = 20 m/min, feed rate f = 0.04 mm/rev, laser power P = 600–2600W.

The characteristics of applied edges are shown in **Table 2**.

	Type of edge material	Symbol of insert material	Coating	Insert
1	Sintered carbide inserts with fine-grain substrate	KC5510	(PVD) TiAlN	SNMG120408
2	Sintered carbide	H10S	Cemented Uncoated	SNMG120408
3	Oxide ceramics ($Al_2O_3 + ZrO_2$)	AC5	Cemented Uncoated	SNGN 120708T 02020
4	Mixed ceramics (Al_2O_3 + TiCN)	MC2	Cemented Uncoated	SNGN 120708T 02020
5	Polycrystalline diamond PCD	KD100	Cemented Uncoated	TPGN110304F

Table 2. Characteristics of applied edges.

3. Results

3.1. Laser surface heating

In **Figure 5**, an example of two point measurements (θ_0 and θ_2) for Si_3N_4 ceramic is presented. The slow heating process of rotating Si_3N_4 ceramics up to required and accepted θ_2 temperature into control system was done in t = 0- to time interval only. The system control unit was increasing laser power irradiation after switching on the v_f feed motion. The laser power irradiation and cutting tool were displaced; therefore, the laser power increase was necessary to get θ_2 constant temperature of the cutting tool point on the machined surface. As a result, the θ_0 temperature increase was noticed in the first phase. The explanation of this phenomenon

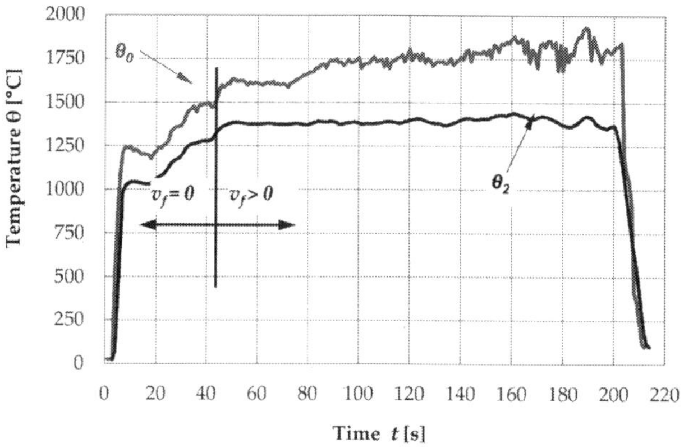

Figure 5. Temperature distributions of the θ_0 and θ_2 during laser-assisted turning within temperature control.

http://dx.doi.org/10.5772/63417

is low thermal conductivity of heated ceramics; therefore, the first phase heating time during feed on required higher laser power irradiation level.

In **Figures 6** and **7** the courses of MMC surface temperature during heating with turning kinematics are shown. Temperature in heated area by laser beam succumb to stabilization (**Figure 7**), varying with the increase of temperature in time on *B* area (**Figures 6** and **7**). This is due to the heat accumulation in the examined sample. Range of temperature Θ_1 equals about 30°C (**Figure 6**), varying based on the different thickness of absorption layer (gouache) on sample surface as well as differences of surface texture for example surface roughness. There is noticeably varied increase of temperature $\Theta_1$from 223 to 510°C (**Figure 7**) in plan machining area (*B*).

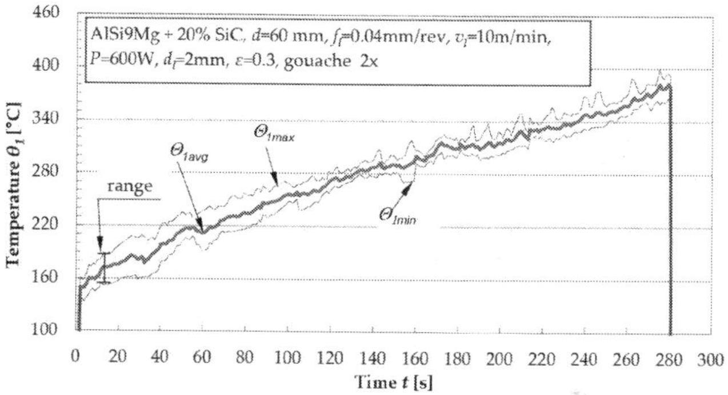

Figure 6. Courses of temperature Θ_1 (*B* zone) during heated MMC by laser beam with maximum and minimum values.

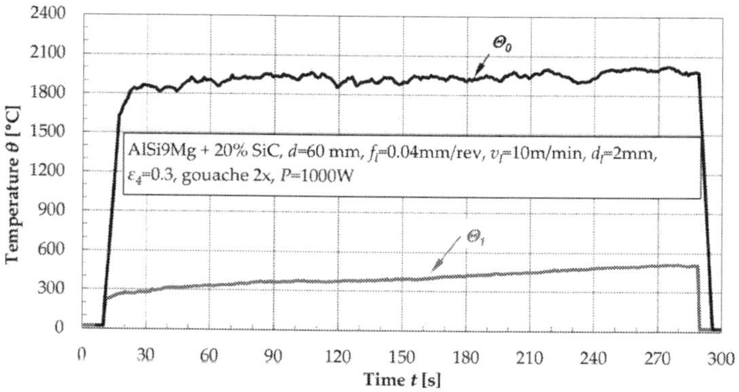

Figure 7. Exemplary courses of temperature Θ_1 in *A* (Θ_0) and *B* (Θ_1) areas during laser heating metal matrix composite.

The **Figure 8** shows the temperature course in time for cemented carbide sample, close to the place of the incidence of the laser beam for different laser powers with heating speed of

20 m/min. The courses of the temperature measurement are stable and do not increase over time.

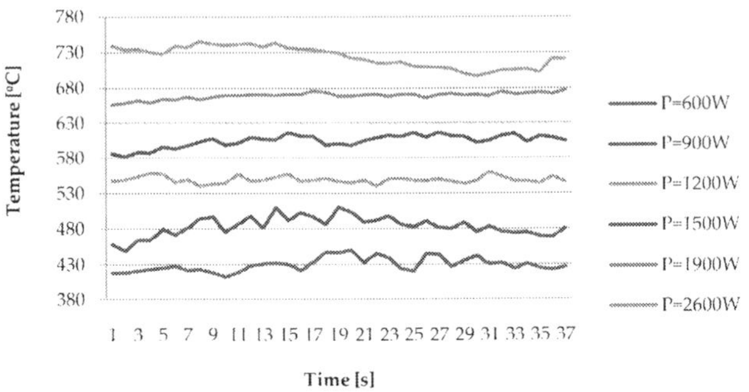

Figure 8. The temperature profile over time for cemented carbide with different laser power in place of the heating area v_l= 20 m/min.

3.2. Tool wear

In **Figure 9**, the investigation results are shown. The laser heated machined ceramic surface temperature increase has decreasing effect on the **VB**$_c$ wear insert tool indicator and the lowest values of it for θ_2 = 1400–1500°C temperatures were noticed. The value of wear inserts tool in those temperatures is more than 5 times lower than wear without using laser heating of machined layer. The inserts tool abrasive and adhesive wear symptoms were observed in the range of investigated temperatures. However, thermal wear symptoms were not apparent. The θ_2 = 1400°C as an optimal temperature was selected for further investigation of machined surface laser heating, because the selection of θ_0 > 2000°C results in an evaporation risk on machined material.

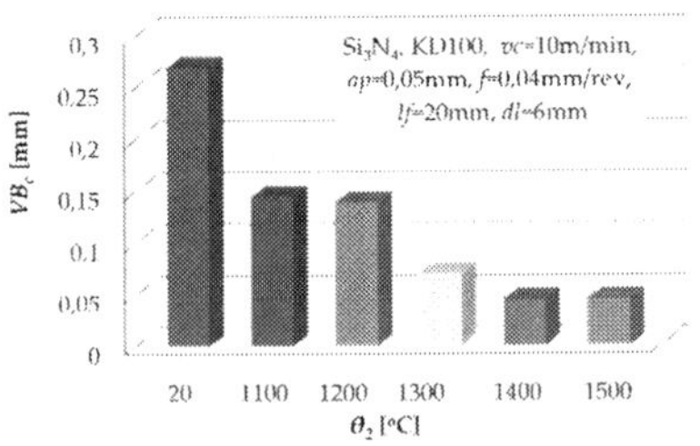

Figure 9. Polycrystalline diamond (PCD) inserts wear in function of the machined surface temperature after 1 tool pass.

In **Figure 10**, wear distributions of investigated tool wedges during laser-assisted turning of ceramic process for selected temperature are presented. The value of the tool insert wear indicator in the range of 1300 and 1400°C temperatures is diversified for the first few passes. However after the next few passes, the difference becomes smaller and after 10 minutes time of machining is insignificant. It could be expected that in real-time analysis the presented wear plots below will look a little bit different, because high temperature exposure of the tool wedges will be much longer than investigated time interval.

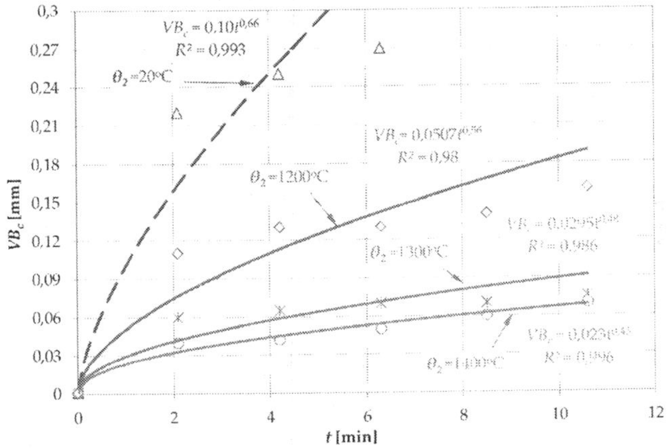

Figure 10. Polycrystalline diamond (PCD) wedges wear distributions during laser-assisted turning process for different Si_3N_4 machined surface temperature. The applied parameters: v_c = 10 m/min, a_p = 0.05 mm, f = 0.04 mm/rev.

The results of the influence of the heating temperature (θ_2=1400°C) cut layer of ceramic Si_3N_4 on tool wear for different wedges are shown in **Figure 11**. It can be observed that the lowest value of tool wear was for PCD and the largest for ceramics wedges.

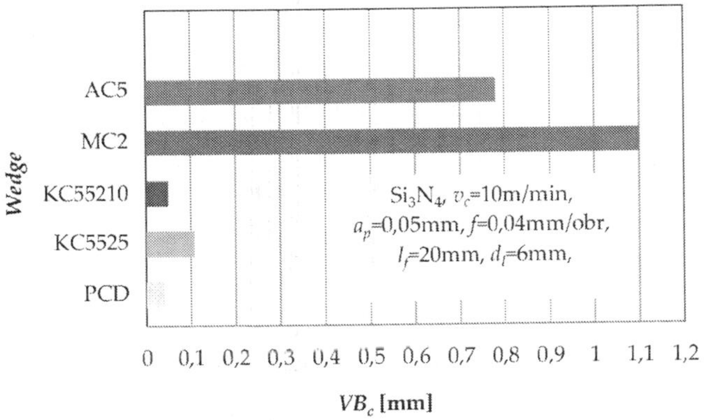

Figure 11. Tool wear for different cutting wedges.

Wedge wear of the AC5 in temperature 1300°C was minimum and at three times lower than at room temperature (**Figure 12**). Tool wear of ceramics as compared to the cemented carbide was very large and the wedges are classified as unable to withstand such difficult conditions.

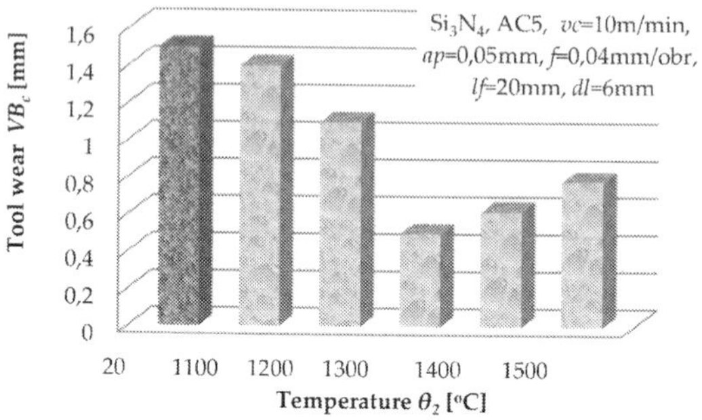

Figure 12. Ceramics AC5 wedge wear in function of machining of the machined surface temperature Θ2.

Figure 13 shows the influence of LAM on the flank wear during turning of Al-SiC metal matrix composites. These results indicated that as a consequence of heating the wedge wear was decreased significantly for tested tools. In **Figure 13**, it can be observed that for LAM, wedge wear is about 31% lower in comparison to conventional turning with cemented carbide and 35% for polycrystalline diamond wedge. However, it is equally important to note that when comparing all results, a laser power of 1000W is the most likely reason giving the optimum tool wear when machining Al/SiC metal matrix composite. The research confirmed the thesis that the softening of the matrix material composite allows pushing in or sliding hard reinforcing particles in the surface layer thus reducing the wear of cutting wedges.

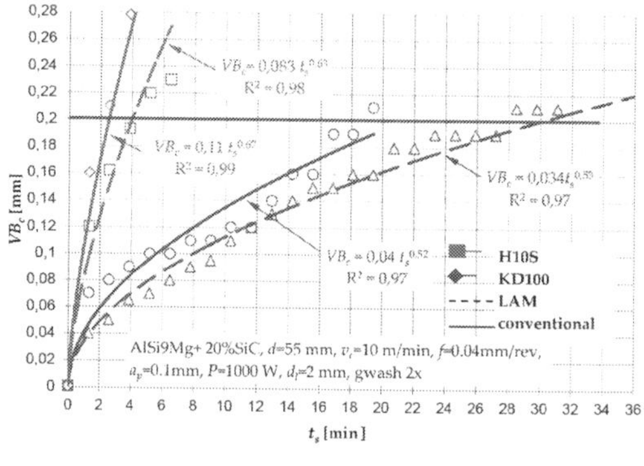

Figure 13. The evolution of the tool wear VBc indicator during turning the metal matrix composites with different feed rate values, the diamond and cemented carbide wedges.

The main wear pattern was observed as regular flank wear. Flank wear is due to the abrasive action of the reinforced particles presence in the MMC. The hard particles SiC of hardness about 2600HV grind the flank face of the cutting tools in similar way as a grinding wheel during machining of materials.

3.3. Machined surface roughness

One of the important indexes of finished product is surface quality. For these experiments, average machined surface roughness values of LAM were found to be lower than those of the conventional turning for all tested tools.

Machined surface roughness of Si_3N_4 ceramics after turning with polycrystalline diamond wedges (PCD) indicates some regularities (**Figure 14**). The increase of temperature in cutting layer (in majority of reported cases) increases the height of roughness profile. The increase of tool wear at the different cutting temperatures does not change this regularity. Therefore a thesis can be stated, that microroughness surface height is not related to the cutter's microgeometry, but temperature of cutting layer in laser-assisted process.

Similar to courses described earlier were those obtained for the turning with cemented carbide wedges KC5510 (**Figure 16**).

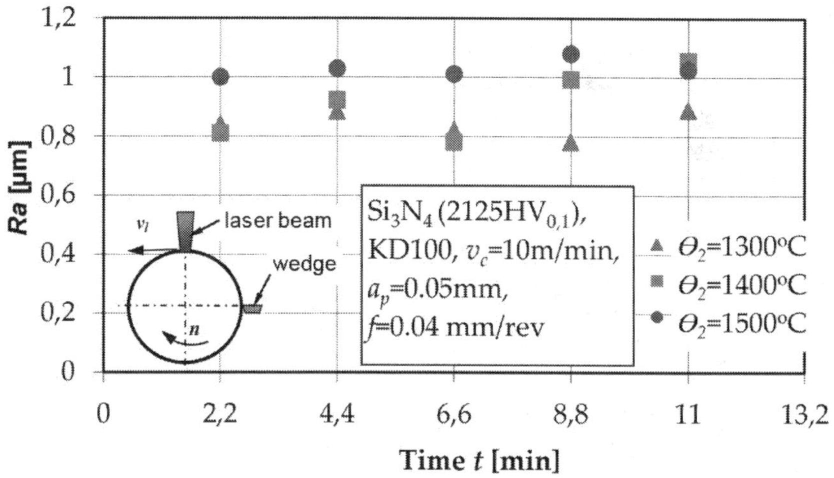

Figure 14. Machined surface roughness in function of machined Si_3N_4 surface temperature, after turning with PCD polycrystalline diamond wedges.

(for $t_s \approx 2$ min) Si_3N_4ceramic machined surface, it can be noticed that in the analyzed temperature range of cutting layer, surface roughness after turning with carbide KC5510 cutter is higher than that generated for the PCD wedge (**Figure 15**).

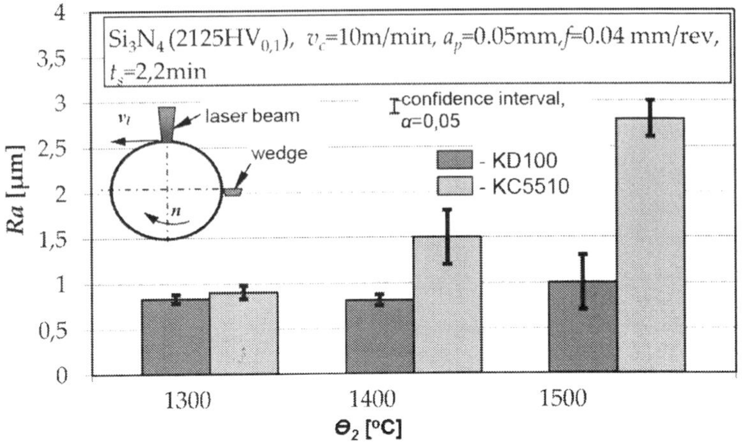

Figure 15. The initial roughness (ts ≈ 2 min) of machined surface for difference temperatures of machining layer after PCD and KC5510 wedges turning.

The increase of the machined surface roughness of Si_3N_4 ceramics, together with the temperature growth, can be explained by changes in the decohesion mechanism of the workpiece. Lower temperatures of the process are dominated by a brittle cracking, while the higher ones are dominated by a plastic flow of work material. Major role in the creation formation of machined surface during turning with a laser heating of a cutting layer have microchips with tendency to adhere to workpiece, and thus to increase in the height of microroughness.

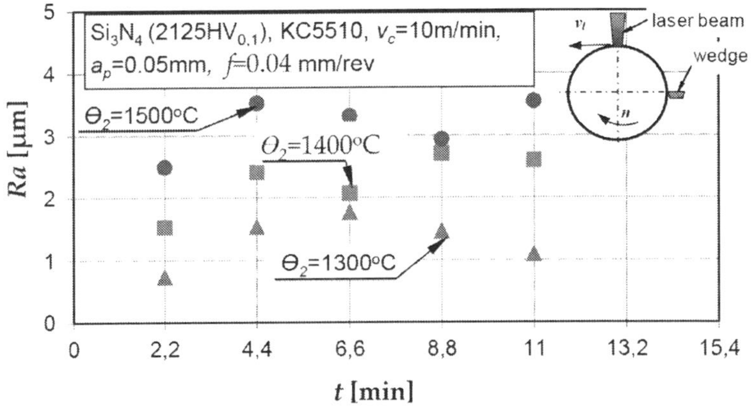

Figure 16. Machined surface roughness distributions during laser-assisted turning process for different machined surface Si_3N_4 temperature, after turning with KC5510 carbide wedges.

The higher temperature of ceramic materials in machining layer during the LAM process improves the cutting ability evaluated by a tool wear [11] but has an adverse effect on quality of machined surface.

Average width of the profile roughness grooves (*RSm*) are, in the investigated range, higher than the feed rate values (**Figure 17**). This indicates difficulties in formation mechanism of the machined surface, inducing serious disturbances in kinematic-geometric projection of the cutter into the workpiece.

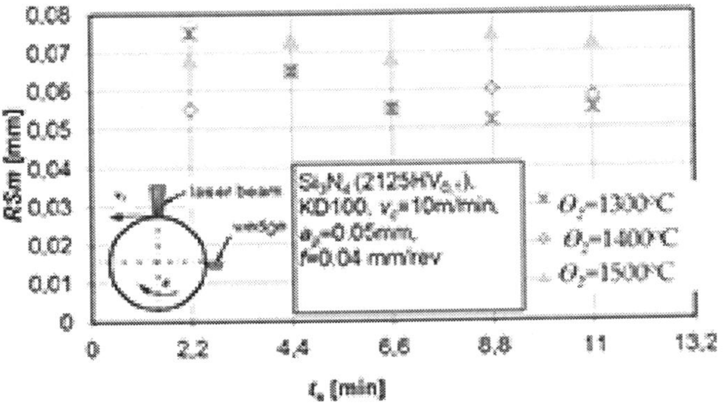

Figure 17. Machined surface roughness RS_m parameter distributions for the different machined Si_3N_4 surface temperature, after turning with PCD wedges.

Figure 18 shows that surface texture has a random character without any wedge mapping in the investigated feed range (f = 0.04 mm/rev). However, the graph of 3D profile (**Figure 18a**) and power spectral density (PSD) graph reveal that dominant peak is different from the applied feed rate (**Figure 18b**).

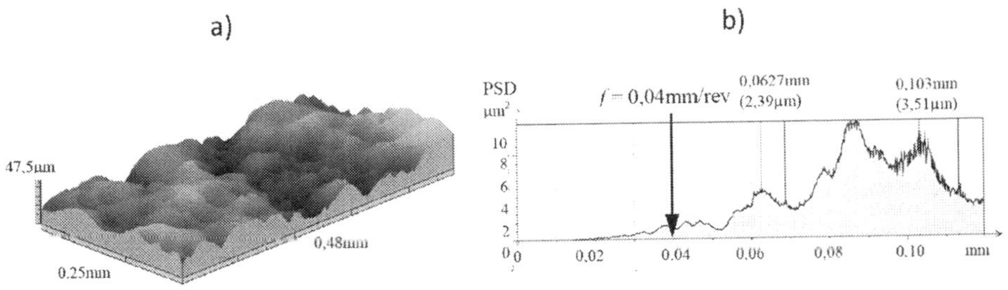

Figure 18. View of laser-assisted turning Si_3N_4 ceramics by polycrystalline diamond (PCD) wedges. (a) surface texture and (b) power spectrum density of the roughness profile. The applied parameters: θ_2 = 1400°C, v_c = 10 m/min, a_p = 0.05 mm, f = 0.04 mm/rev.

Nevertheless, in case of turning with KC5510, clear microstructure's orientation can be seen (**Figure 19**), which is attributed to the kinematic-geometric projection of the cutter into the workpiece. This observation is confirmed by a PSD diagram (**Figure 19b**), where the modal value of profile component has a frequency close to the feed rate.

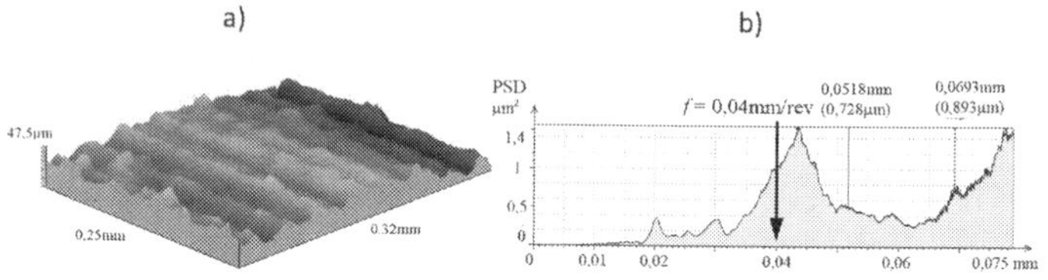

Figure 19. View of Si_3N_4 ceramics surface texture (a) and power spectrum density of the roughness profile (b) after laser-assisted turning within carbide (KC5510) wedges. Parameters: θ_2 = 1400°C, v_c = 10m/min, a_p = 0.05mm, f = 0.04mm/rev.

The machined surface roughness is one of the most important quality indicators of the MMC materials. **Figure 20** shows the influence of LAM on the machined surface roughness *Ra* parameter, during turning of Al/SiC metal matrix composite with different cutting wedges. Significant differences in the roughness of machined surface were observed for the laser-assisted and conventional turning (*P*= 0 W) with the carbide wedges. However, for the conventional turning, it is seen that average machined surface roughness value is higher than that obtained for the LAM. It could be explained by the fact that, in conventional turning, SiC particles were pushed through the soft matrix or pushed out of matrix during machining. Consequently, cracks and pits were formed on the machined surface (**Figure 10**), which are inducing poor surface finish. During hybrid machining, laser heated layer is being cut in the liquid or semi-liquid state, so that fills the grooves in the machined surface generated after plowing or crumbling out the reinforcement particles. This mechanism reduces the average height of machined surface roughness.

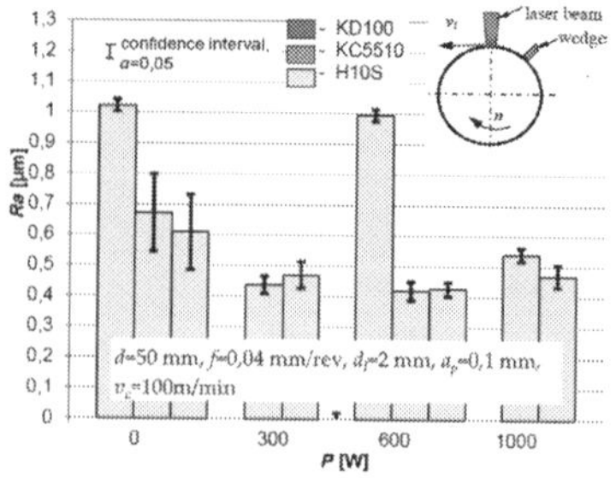

Figure 20. The influence of a laser's beam power on machined surface roughness *Ra* after LAM and conventional turning of aluminum-ceramic via different cutting wedges material.

http://dx.doi.org/10.5772/63417

The analysis of power spectrum density (PSD) of the roughness profile revealed that machined surface texture has random character with the significant disturbances of the cutting edge kinematic-geometric projection. The PSD chart has dominant component different from the applied feed rate. The analysis of power spectrum density for different values of feeds (f_1 = 0.07 mm/rev, f_2 = 0.1 mm/rev) reveals significant disturbances of the cutting edge kinematic-geometric projection (**Figure 21**).

In **Figure 21b**, the orientation of surface texture was observed as a result of turning kinematics; however, a power spectrum density diagram did not reveal a component related to the employed feed rate. **Figure 21a** and **21b** shows a cavity created after crumbling out the SiC particles from metal matrix.

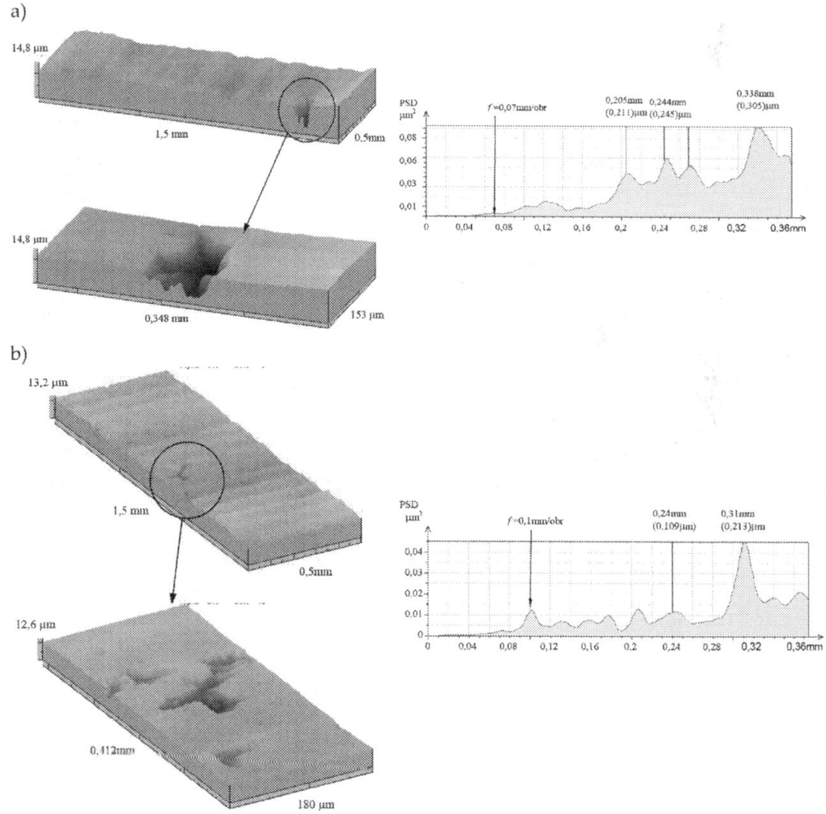

Figure 21. 3D topographic images (machined surface of metal matrix composite and power spectrum density) of surface roughness profiles after conventional turning of non-coated carbide H10S wedge for feed rate: (a) f = 0.07 mm/rev, (b) f = 0.1 mm/rev.

However, in the case of conventional turning, there was an increase in average surface roughness for analyzed sample in comparison with LAM. The formation of grooves on the sample as a result of moving of SiC particles by cutting tool is a major cause of the deterioration of surface roughness with traditional turning. With a LAM the temperature in cutting zone increased and leads to weakening the connections between the SiC particles and aluminium matrix. In most cases, it causes pushing the particle outside or in to the aluminium alloy. The same observation had Altinkok [16] in research.

Figure 22 shows a comparison of the surface profiles of the machined surface by conventional cutting and LAM. The profiles show the difference of surface finish achieved.

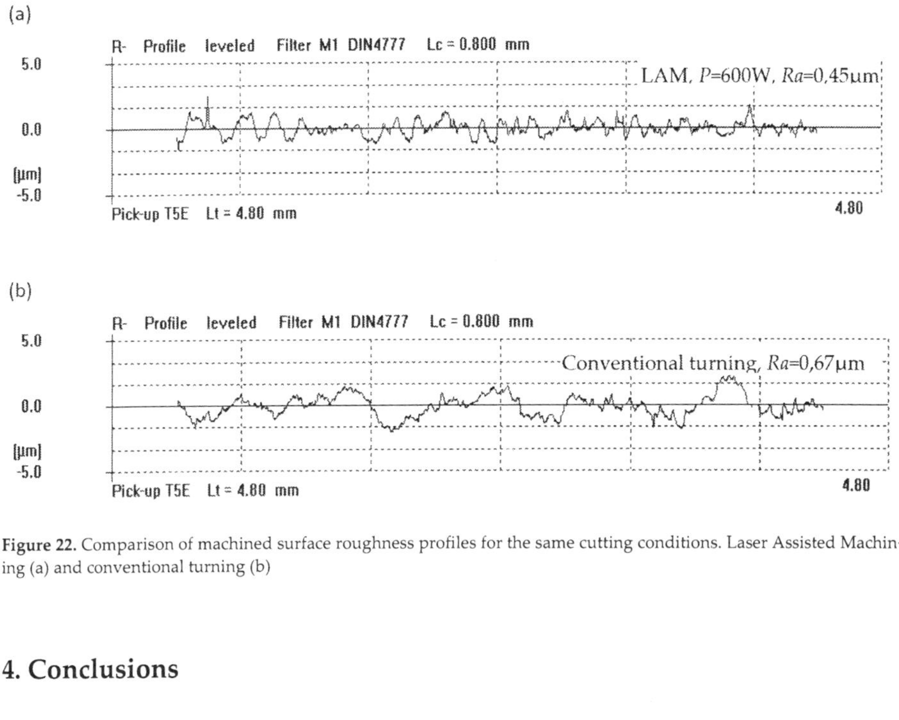

Figure 22. Comparison of machined surface roughness profiles for the same cutting conditions. Laser Assisted Machining (a) and conventional turning (b)

4. Conclusions

Laser enhanced conventional turning to heat the cutting materials, as well as to change the microstructure or locally harden the material in front of the cutting tool. This process is carried out in order to facilitate the machining due to its softening and change of the workpiece's deformation behavior. The local temperature of the material in the shear deformation zone plays an important role in the thermally enhanced machining process.

The investigations on laser-assisted turning of difficult to cut materials confirmed that laser heating of the machined surface causes significant improvement in turning process, which is estimated by tool wear and surface roughness indicators.

The low tool wear intensity in a suitable temperature is crucial for increasing the tool life. This opportunity enables the significant reduction of cost tool production.

The flank face wear was observed during laser-assisted and conventional turning. The abrasive tool wear mechanism is dominant in turning of these materials. Turning with laser heating improves machined surface roughness in comparison with conventional turning for all examined wedges.

Acknowledgements

The author wishes to gratefully acknowledge Mr. Marian Jakowiak for valuable advice, support, time, and available materials that contributed in creating the work. The part of presented research results, executed under the domestic project LIDER of No 141/L-5/2013, was funded with grants for education allocated by the National Centre for Research and Development

Author details

Damian Przestacki

Address all correspondence to: damian.przestacki@put.poznan.pl

Poznan University of Technology, Poznan, Poland

References

[1] Przestacki D., Jankowiak M. Surface roughness analysis after laser assisted machining of hard to cut materials. Journal of Physics: Conference Series. 2014; 483: 012019, Published online: 28 March 2014.

[2] Przestacki D. Conventional and laser assisted machining of composite A359/20SiCp. Procedia CIRP. 2014; 14: 229–233.

[3] Davim J. P., Baptista A. M. Relationship between cutting force and PCD cutting tool wear in machining silicon carbide reinforced aluminium, Journal of Materials Processing Technology. 2000; 103: 417–423.

[4] Naher S., Brabazon D., Looney L. Computational and experimental analysis of particulate distribution during Al–SiC MMC fabrication. Composites: Part A. 38; 2007; 38: 719–729

[5] Bassani P., Capello E., Colombo D., Previtali B., Vedani M. Effect of process parameters on bead properties of A359/SiC MMCs welded by laser. Composites: Part A. 2007; 38: 1089–1098.

[6] Davim J. P. Diamond tool performance in machining metal-matrix composites. Journal of Materials Processing Technology. 2002; 128: 100–105.

[7] Skvarenina S., Shin Y. C. Laser-assisted machining of compacted graphite iron. Journal of Machine Tools and Manufacture. 2006; 46: 7–17.

[8] Manna A., Bhattacharyya B. A study on different tooling systems during machining of Al/SiC-MMC. Journal of Materials Processing Technology. 2002; 123: 476–482.

[9] Pramanik A., Zhang L. C., Arsecularatne J. A. Machining of metal matrix composites: effect of ceramic particles on residual stress, surface roughness and chip formation. International Journal of Machine Tools and Manufacture. 2008; 48: 1613–1625.

[10] Chang C. W., Kuo C. P. Evaluation of surface roughness in laser-assisted machining of aluminum oxide ceramics with Taguchi method. Machine Tools and Manufacture. 2007; 47: 141–147.

[11] Jankowiak M., Bartkowiak K. Machinability of laser heated silicon nitride ceramics during turning process. Proceedings of the 25th International Congress on Applications of Lasers & Electro-Optics ICALEO, Scottsdale, AZ, USA. 2006; 311–316.

[12] Ding H., Shin Y. C. Laser-assisted machining of hardened steel parts with surface integrity analysis. International Journal of Machine Tools & Manufacture. 2010; 50: 106–114.

[13] Ding H., Shen N., Shin Y. C. Thermal and mechanical modeling analysis of laser-assisted micro-milling of difficult-to-machine alloys. Journal of Materials Processing Technology. 2012; 212: 601–613.

[14] Sun S., Brandt M., Dargusch M. S. Thermally enhanced machining of hard-to-machine materials-a review. International Journal of Machine Tools & Manufacture. 2010; 50: 663–680.

[15] Dandekar C. R., Shin Y. C. Multi-scale modeling to predict sub-surface damage applied to laser-assisted machining of a particulate reinforced metal matrix composite. Journal of Materials Processing Technology. 2013; 213: 153–160.

[16] Altinkok N. Investigation of mechanical and machinability properties of Al2O3/SiCp reinforced Al-based composite fabricated by stir cast technique. Journal of Porous Materials. 2015; 22 (6) DOI: 10.1007/s10934-015-0048-0.

Chapter 5

DLC Thin Films and Carbon Nanocomposite Growth by Thermionic Vacuum Arc (TVA) Technology

Rodica Vladoiu , Corneliu Porosnicu ,
Aurelia Mandes , Ionut Jepu , Virginia Dinca ,
Aurelian Marcu , Mihail Lungu , Gabriel Prodan and
Liga Avotina

Additional information is available at the end of the chapter

http://dx.doi.org/10.5772/63367

Abstract

The aim of this chapter is to report the results on synthesis **DLC thin films and carbon nanocomposites by** the versatile **nanofabrication method based on plasma** entitled **thermionic vacuum Arc (TVA)**. TVA technology is based on the localized ignition of the arc plasma in vacuum conditions. Among thin film coating methods by vacuum deposition techniques with high purity, low roughness, and good adhesion on the substrates, **TVA is one of the major suitable methods to become a powerful coating technology**. Two or three different TVA discharges can be ignited simultaneously in the same chamber for multi-material processing using TVA and separate power supplies. These TVA discharges are localized and do not interfere with each other. Simultaneous two or three TVA discharges were already used for the production of alloy/composite of various materials. This is due to the high versatility concerning the configuration of experimental arrangements, taking into account the number of electron guns, symmetry of the electrodes, relative position of the anode versus cathode, and also the huge opportunity to combine the materials to be deposited: bi- and multi-layers, nanocomposites, or alloys in order to have specific applications. This chapter presents the comparative results concerning the surface-free energy information processing, the reflective index, the hardness, and the morphology to provide a coherent description of the diamond-like carbon films and carbon nanocomposites synthesized by thermionic vacuum arc (TVA) and related configurations where Me = Ag, Al, Cu, Ni, and Ti: **binary composites (C-Me, C-Si) and ternary composites (C+Si+Me)**. The results include reports on the distribution in size, surface, geometry, and dispersion of the nanosized constituents, tailoring and understanding the role of interfaces between structurally or chemically dissimilar phases on bulk properties, as well as the study of physical properties of nanocomposites (structural, chemical, mechanical, tribological). The results presented here could have a great impact on the development of advanced materials and many manufacturing industries, as well as expanding the technologically important field of interface science where the control of the film-substrate interface would be critical.

Keywords: DLC, nanocomposites, carbon, thermionic vacuum arc, nanofabrication

1. Introduction

Diamond-like carbon (DLC) is an important form of amorphous carbon consisting of a mixture of both sp^2 and sp^3 coordinated carbon. It is also known that hydrogenated carbon films show a wide range of properties concerned with hardness (3000–5000 kg mm^{-2}), chemical inertness, thermal conductivity, wear resistance (1.6 x 10^{-9} mm^3/mN), low friction coefficients (0.1–0.2), electrical resistivity, optical transparency from ultraviolet to infrared, and tunable band gap by manipulating sp^2 and sp^3 bonding ratio [1–6].

Unfortunately, the widespread use of DLC thin films has been limited by poor adhesion of these films to substrates, because of large compressive stresses induced in the films during the deposition process. It has been reported that the residual stresses in these films arise from a change in the local chemical bonding and changes in bonding angle and bonding length, caused by the ions penetrating into the film surface during processing [3, 7, 8].

The study of multiphase nanocomposite material involves that at least one of the constituent phases has one dimension less than 100 nm. The nanocomposite films are sought due to their multifunctionality, the opportunity of designing unique combinations of properties, difficult to achieve otherwise with traditional materials. The challenges in reaching this promise are tremendous especially in the case of **carbon-based nanocomposites** [9–11].

Thin film depositions on specific substrates can totally alter the physical-chemical properties of the coated material, adding certain advantages. During the operation time of mechanical devices with moving parts, it is desirable to reduce friction between metallic or nonmetallic surfaces. The small variation of the friction coefficient both along the contact surface and during the operation process was attained. High imperviousness to temperatures in the range −200 to 450°C or even on broader domains, high resistance to dust particles and to corrosive action of active chemicals such as acids, bases, and the possibility to work without lubricants for a limited period of time in emergency cases are some of the properties of the obtained coatings. These results were confirmed by specific analyses [12, 13]. One of the goals accomplished with the use of these composite materials was to protect the devices against the damaging action of water and vapors thus increasing their operating lifetime. Also, an enhancement in the appearance of decorative parts was observed. Ternary deposition of these materials can reduce the surface wearing caused by the contact erosion, gripping, etc. and also helps avoiding the stick-slips. Also, the creaking between relatively moving parts was completely removed.

Due to the deposition process, the obtained coated part has a great resistance to applied pressure. From an economic point of view, the part production cost is reduced significantly due to the fact that the main body can be made of a bulk material with small fabrication cost on which an insignificant quantity of composite layers is deposited. These depositions can be used to enhance the characteristics of mechanical parts such as automotive parts, springs, arbors, gears, cranked axles, central pins, bearings, etc. These specific coatings must have certain characteristics such as resistance to abrasion, very good adherence to substrate, compatibility with the substrate both physically and chemically with no destructive attack of the substrate, lack of friction electric charge generation in order to avoid sparks, and explosions in organic chemical vapors [14]. An important aspect was related to the high wear resistance of the deposited materials. Materials used for the deposition must have the characteristics mentioned above and also present a certain consistency soft enough to allow compatibility with the substrate, to follow its form, and be adherent to it. Thus, the deposited materials used for this study must be both hard and soft.

In this chapter, results on synthesis of **DLC thin films and carbon nanocomposites by** the versatile **nanofabrication method based on plasma** entitled **thermionic vacuum arc (TVA) are reported**.

The Romanian original method, **thermionic vacuum arc (TVA)**, has been studied and successfully used for the synthesis of a large range of metals, oxides, and carbon, as attested by published papers, patents, and internationally funded projects. There is a great potential for this method to become a **powerful thin film technology** for a large range of applications. Nanometer-scaled thin film deposition such as metals, alloys, DLC, and refractory metals such as W, Mo, Ta, Nb, Re, and B and also nanocomposites: binary **C+Me**, as well as ternary composites **$C+Me_1+Me_2$**with low roughness, good smoothness, and low friction coefficient, could be synthesized in different TVA electrode configurations.

2. Basic principle of TVA method and experimental setup

The plasma arc discharge is obtained by the bombardment of electrons emitted by the filament from inside of a Wehnelt cylinder directed toward the anode. The material to be deposited from the crucible is heated until evaporation by those accelerated electrons. An increasing positive DC high voltage is applied on the anode, and consequently, the ions of the materials appear by fast electron neutral collisions. The degree of ionization increases due to the accelerated electrons that continuously vaporize the anode material and due to the collision with neutrals, at a certain value of the voltage, a bright plasma is ignited.

The experimental setup for TVA method allows the simultaneous deposition of different materials, providing the possibility of obtaining multi-component thin films in a special two-electron gun configuration (**Figure 1**) or three-electron gun configuration (**Figure 2**). The electron guns are symmetrically arranged with respect to the holder mounted at an optimum distance on the central axis. The electric arc is ignited between the cathode and the anode material carbon or metal continuously evaporated by the electrons accelerated at high voltage and incident of the anode. The anode temperature of each element was adjusted in order to have comparable evaporation rates.

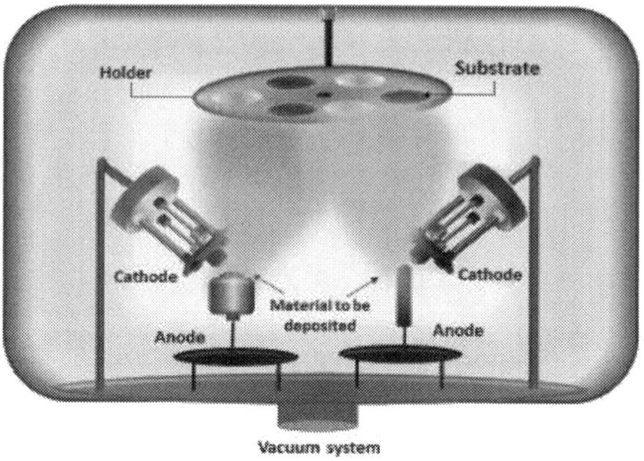

Figure 1. Experimental setup for binary films.

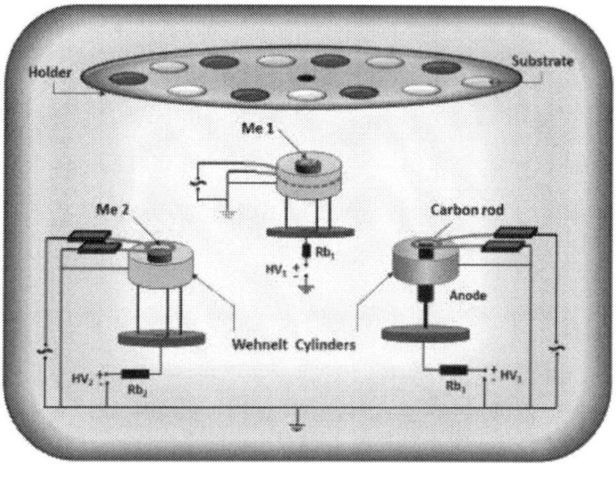

Figure 2. Experimental setup for ternary films.

The originality of the TVA method consists in the fact that the energy introduced into the system for ignition of the plasma is supplied simultaneously by an electron gun and a high-voltage source. The creation of the vapors is obtained by the electron bombardment of the anode material (e.g., carbon or metal) with electrons emitted by the filament and accelerated between the electrodes. The thin film is growing under the bombardment of incident energetic ions, besides the neutral atoms of the material to be deposited from the anode. Taking also into account the discharge performed in vacuum conditions, one of the great advantages of this method consists of its high degree of purity. Another advantage is related to the direct proportionality between the energy of ions and the cathode fall. The cathode is at ground potential so, with respect to the wall of vacuum vessel, the plasma potential is equal to the cathode fall. Therefore, for characteristic anode currents of 1 A, a potential difference equal to the cathode fall will accelerate the ions toward the walls of the vacuum vessel up to energy of 500 eV.

3. DLC thin films

DLC of varying sp^3/sp^2 ratio was obtained by changing the deposition conditions. The sp^3 phase is responsible for the high mechanical properties and the sp^2 phase for the electron conductance properties of DLC films [15, 16]. It was found that in order to tune amorphous carbon to have a specific sp^3 fraction, the following parameters of the TVA technology are essential: filament current (I_f), arc current (I), applied voltage (V), and the distance to the substrate (d). The influence of these plasma parameters on the preferential formation of sp^3 sites was analyzed.

In the case of carbon, the previous work proved that thermionic vacuum arc method can be used to produce smooth DLC films with relatively low growth stress and high nanohardness with a controlled final percentage of sp^3 content, between 20–67%—depending on plasma parameters. It was found that the sp^3:sp^2 ratio is higher at higher applied voltage, when the plasma potential and consequently the carbon ion energy are higher [17].

The extension of thermionic vacuum arc technology to coatings using gases or evaporable liquids as in the case of gaseous thermionic vacuum arc (GTVA) instead of solid materials is of great interest especially for DLC growth. While in the TVA method the anode is solid—a carbon rod—being suitable for practically any solid material, the GTVA method enables the growth of ultrathin and ultrapure DLC coatings using CH_4 as gas precursor in vacuum conditions [18].

Although the principle is the same, there are main operational differences between the two forms of TVA technology (electrodes configuration, breakdown voltage, specific parameters), not pointed out in this chapter. Both type of discharges offer the following advantages for thin carbon layer depositions: high rate of evaporation up to 100 nm/min; no buffer gas is necessary; very low thermal energy transfer and very stable discharge conditions. The prepared thin film layers are smooth, compact, and of very fine structures; no cathode impurities can be found in deposited thin films; low mechanical stress of the deposited layer; and very good adherence.

The deposition of a-C/DLC films have a rather high residual compressive stress which depends on the deposition method. The limit for practical application of a-C/DLC film high stress is caused by the weakening of film bonding to the substrate; thus, when reaching some critical thickness, the stress can result in the delamination of film from the substrate [19, 20].

For this reason, the initial steps have been undertaken in this direction in order to optimize the structure of carbon-metal composite films and the interaction with other materials.

4. Binary composites deposited by TVA technology

It is well known that one way to alter the grain size of the material in a controlled way is by using the co-deposition of two materials. Practically, there is no limit for TVA deposition to combine two elements in order to obtain binary composites. A plenty of mixture based on carbon has been synthesized during the last years by the TVA method [21–24]. We will focus on the best combination for anticorrosion and wear resistance properties: C-Ag, C-Al, C-Cu, C-Ni, and C-Si.

As a soft and ductile element, **Cu**, as well as **Ag**, embedded into the amorphous carbon matrix, may improve tribological properties such as lowering brittleness rates. C-Cu films are considered as high-technology materials due to their unique characteristics that have made them of particular interest in a wide variety of applications.

The excellent properties of the Ag-incorporated carbon films can be related to the following: first, due to the face-centered cubic structured, Ag nanoparticle has a significantly smaller elastic modulus than carbon, which can absorb compressive stress from the carbon matrix [25, 26]. Meanwhile, the formation of bonds between the nanocrystallites and the matrix can be diminished due to an increase in toughness of composite film [27–29].

Also, the C-Ag thin films present a high interest thanks to the prospect of various applications such as catalysts, optical filters, and photographic processes for biology and medicine [30, 31]. The incorporation of silver into carbon films offers the possibility to provide coatings that are hemocompatible and antibacterial [32–35].

Another element **Al** is remarkable for its low density and for its ability to resist corrosion being less affected by environmental factors such as air and water and consequently being suitable for coating application in open air.

The morphology (a), composition (b), and scanning electron microscopy (SEM) images after the tribological tests are as shown in **Figure 3** for all these binary composites synthesized by the TVA method.

In the same context, due to the high resistance to corrosion of **Ni**, it was found that C-Ni coatings on stainless steel were very effective in improving the wear resistance. Surface morphology analyzed using an AFM (Atomic Force Microscope) in tapping mode also shows a significant roughness of the coating. **Figure 4a** shows a typical image of the of a C-Ni sample (Ni concentration of about 40%), and the RMS - Root Mean Square roughness is of about 500 nm. In this case, C-Ni coating seems to have followed the substrate morphology creating a rough surface. However, for friction parameters this is actually beneficial for the retention of the lubricating liquid in the running-in phase or after long time nonoperating mode of the engine shaft.

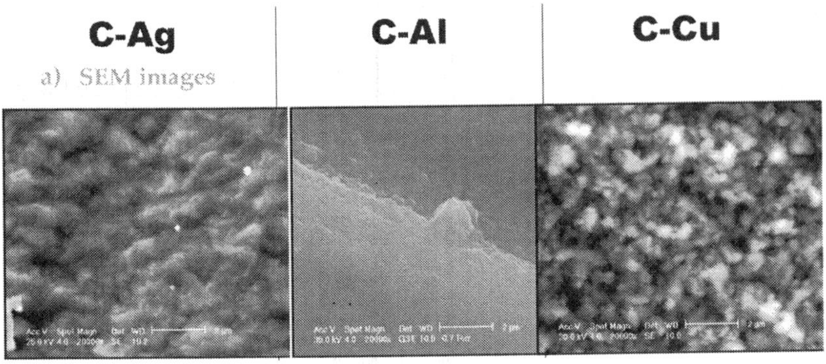

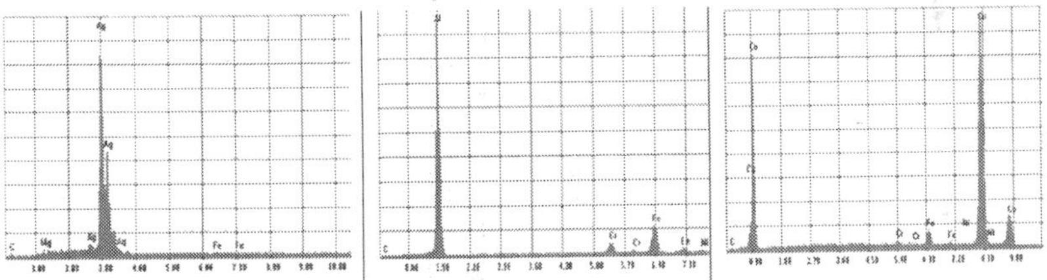

C-Ag:

Element	Wt %
C	2.1
Mg	1.39
Ag	95.8
Fe	0.71
Total	100

C-Al:

Element	Wt %
C	2
Al	74.04
Cr	5.35
Fe	17.14
Ni	1.48
Total	100

C-Cu:

Element	Wt %
C	3.34
Cr	0.86
Fe	2.98
Ni	0.91
Cu	91.92
Total	100

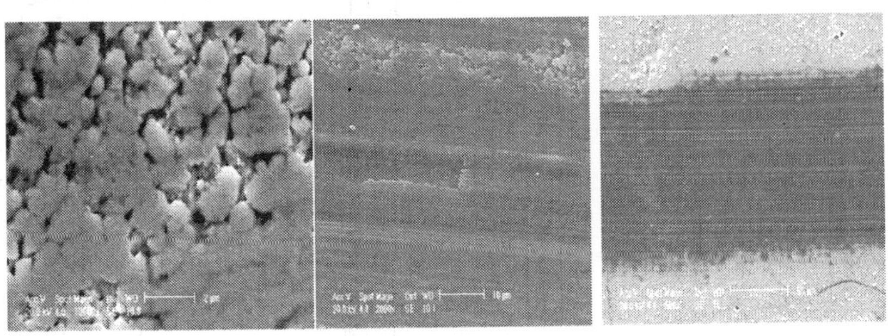

Figure 3. Topographical images and compositional results for **C-Ag, C-Al**, and **C-Cu** films.

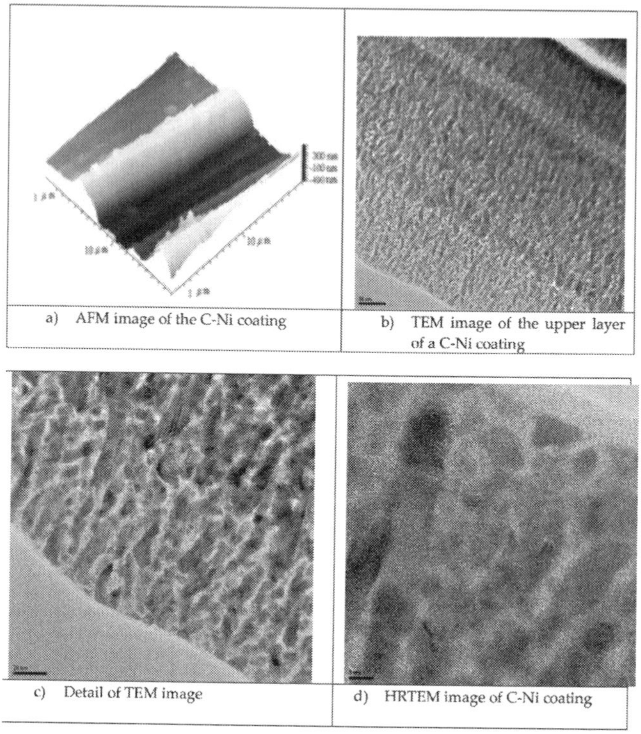

a) AFM image of the C-Ni coating

b) TEM image of the upper layer of a C-Ni coating

c) Detail of TEM image

d) HRTEM image of C-Ni coating

Figure 4. AFM and TEM images of C-Ni coating.

Figure 4b and **c** shows the images from the transmission electron microscopy (TEM) investigations of the C-Ni coatings. The structure of the upper layer presents tubular features with about 10 nm width and 50–100 nm length. These appear together with small grains with a lateral size of about 5 nm.

The high-resolution transmission electron microscopy (HRTEM) image of the grains in the upper layer is presented in **Figure 4d**. One could see the crystalline structure in most of the grains, while the surrounding matrix is amorphous.

Regarding **the frictional properties** of the coatings, the tests were performed using a ball-on-disc tribometer provided by CSM INSTRUMENTS, in ambient atmosphere (~50% humidity and ~23°C). Using a dead weight, a load of 1 N was applied. Carbon-based binary composites were covering 25 mm diameter stainless steel discs, and the counterpart was a 6-mm ball-bearing steel. The track radii were 3 and 4 mm, respectively, and the sliding speed was about 5 mm/s. A total of approximately 10 m of sliding was performed in about 30 min. Recorded friction coefficient (on a PC) was studied using OM (Optical Microscope). As shown in **Figure**

5, the coefficient of friction of the prepared films drastically decreased about three to five times compared to those of the initial substrate (stainless steel) or with pure Ni coatings.

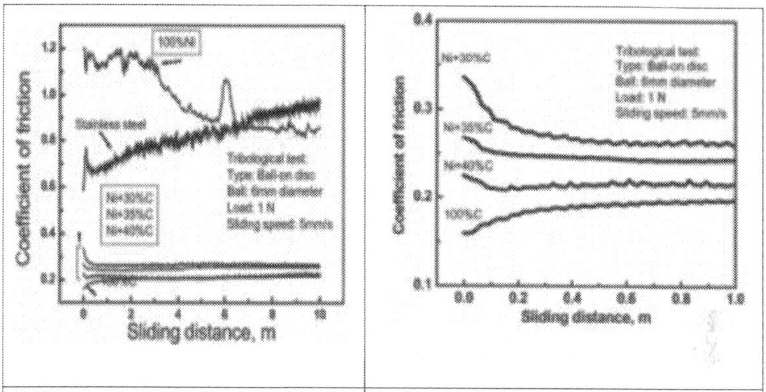

a) The results of the ball-on-disc tests for C-Ni coating	b) Running-in phase results of tribological test for C-Ni coating

Figure 5. Frictional behavior of C-Ni coatings.

Due to the ion bombardment in the TVA technique, compact C-Ni films with no columnar structure were formed on stainless still surface, while irregular and hemisphere-type protuberance C-Cu films were obtained on the same substrate surfaces. In principle, the surface roughness could be further optimized by adjusting the deposition parameters (such as the intensity of the filament heating current, the introduced electrical power and the anode-cathode distance, and the distance between the anodes and the substrate). Measured coefficients of friction of the C-Me (for Me concentration in the range of 30–40%) by using a CSM pin-on-disc tribometer in dry sliding overlays were found to be reduced by a factor of 3 to 5, compared with that of the stainless steel used as substrate or covered with a pure metal [36].

5. Ternary composites deposited by TVA technology

Using the thermionic vacuum arc deposition method, ternary thin film structures with improved physical properties for industrial applications were obtained for the first time. Properties such as resistance to abrasion, good adherence, high hardness, low friction, and a strong compatibility with the coated material were observed. In order to obtain these characteristics, a mixture of C, Al, and Si was deposited on different substrate glass, silicon, and OLC45. **Carbon** was used in this mixture for its high lubrication of the composite layer and high resistance against chemical and corrosive environment attack. In combination with **aluminum**, it ensured the lubrication and a low-friction coefficient value of the thin film.

On the contrary, **silicon** is a very attractive material: **tough, with low friction, second only to diamond in wear resistance**. For this reason, C-Si has excellent properties such as high strength and hardness, thermal and chemical stability, oxidation resistance, high melting point, high erosion resistance, etc. All of these properties make CSi a perfect material for high power, high temperature electronic devices as well as cutting and abrasion applications. Silicon carbide is composed of tetrahedral carbon and silicon atoms with strong bonds in the crystal lattice [37].

The deposition took place under high vacuum conditions. During this phase, the parts were rotated in order to assure a uniform coating. Both carbon and silicon were deposited using the TVA method [38] by igniting a plasma in their pure vapors. The third material aluminum was deposited using a thermal evaporation setup. This setup allows for a smoother control of the deposition rate than in the case of the TVA setup for this material. One of the advantages offered by TVA technique consists in using punctual evaporation sources which allow for a large variety of elemental concentration for ternary depositions by positioning the samples in respect with the sources [39]. The lack of any gaseous inclusion inside the deposition layer due to extremely high vacuum conditions is another advantage offered by TVA deposition method [40]. All these aspects mentioned above are essential for obtaining high-quality depositions.

Figure 6. SEM image of the C-Si-Al film deposited on silicon (left) substrate and wear image (right).

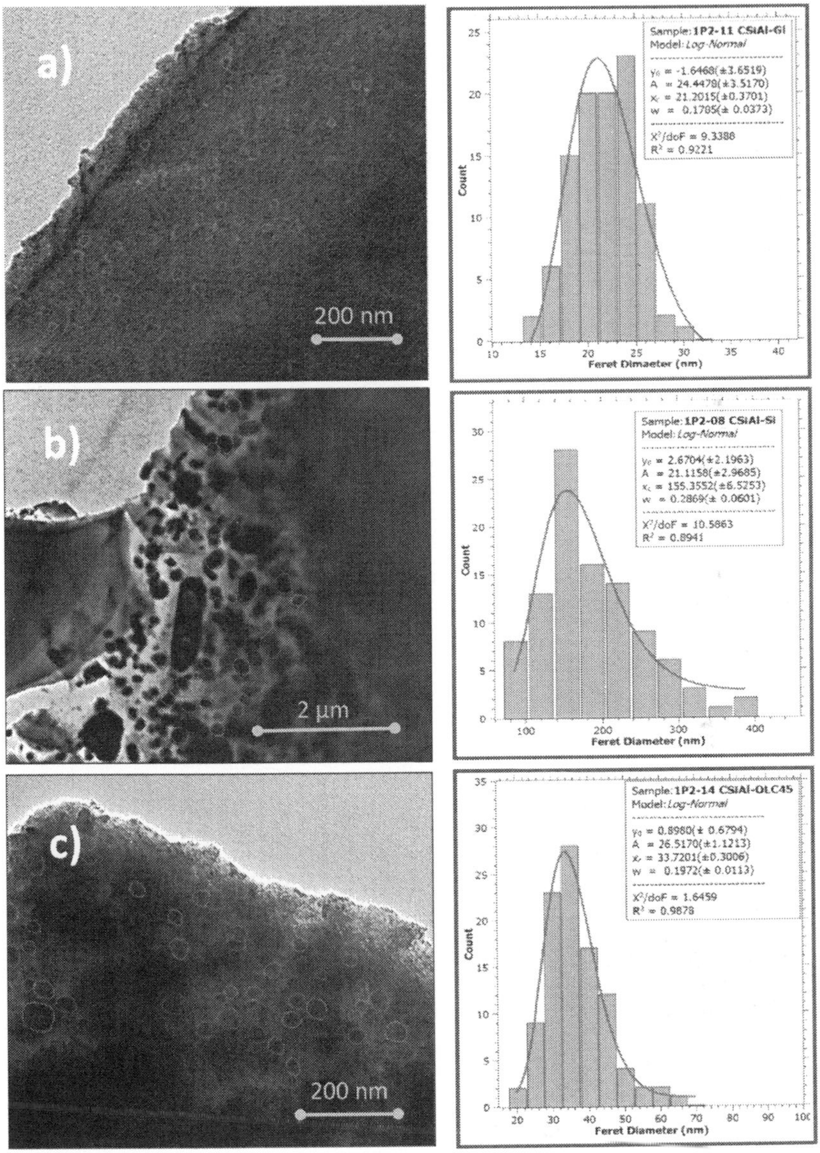

Figure 7. TEM images (left) with particles selection for statistic determination and grain size distribution (right) of the deposited films on different substrates.

SEM and EDS measurements were performed to observe the surface morphology and atomic relative concentration distribution of the three elements inside the film. The SEM image revealed the granular nanocomposition of the film (**Figure 6**). In spite of the fact that EDS (Energy Dispersive Spectroscopy) investigation does not offer information on depth profile since it incorporates the signal of the strained electrons, it can offer exact information on structure. The relative concentrations of the film components determined by EDS investigation were estimated at 58% C, 9% Al, and 30% Si. The increased C content in the film was obtained due to the chosen deposition parameters. The oxygen concentration was estimated to be 2%. This contamination was attributed to the post-deposition exposure of the samples to atmospheric air. We can argue that EDS is not a reliable method for determining silicon content of the C-Si-Al films due to the fact that the high silicon concentration could be attributed to the substrate.

TEM images (**Figure 7**) were fitted with radial distribution functions, and so the peaks were obtained Using log-normal function for the polycrystalline materials. Using the values obtained at some selected crystallites, the histogram is performed for Feret-like diameters.

TEM analysis confirmed an amorphous-like structure for the samples coated on glass and OLC45 substrates, and a crystalline structure for the samples coated on silicon, certifying the role of the substrate structure on the deposition atomic arrangement.

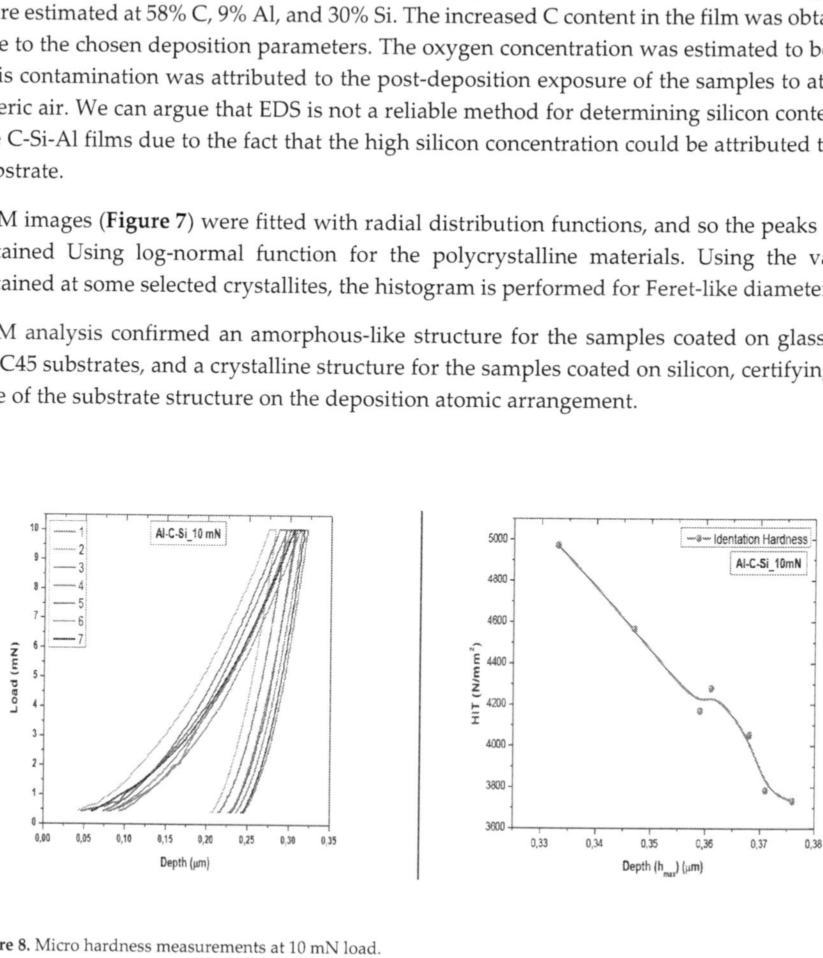

Figure 8. Micro hardness measurements at 10 mN load.

The film hardness was determined by nanoindentation measurements. **Figure 8** (right) shows the result of seven measurements performed on C-Si-Al using a 10-mN force at different depths [**Figure 8** (left)]. Thus, film hardness is very high with values between 3600 and 5000 N/mm^2 (~3.6–5.0 GPa).

http://dx.doi.org/10.5772/63367

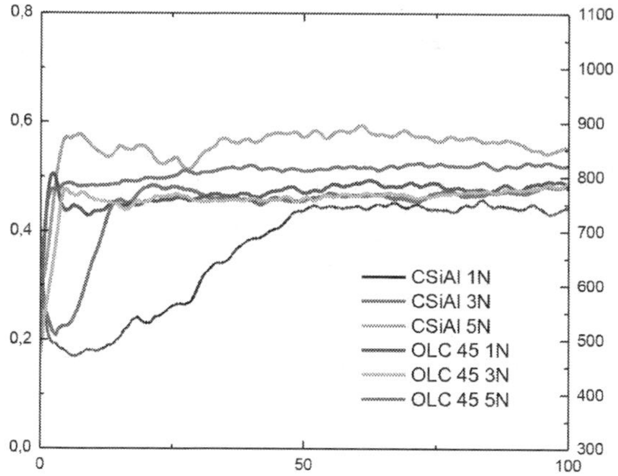

Figure 9. Comparative view of the friction coefficient of substrate (OLC45) and of CSiAl film deposited on the substrate at different loading forces.

For tribological characterization, systematic measurements were performed using a ball-on-disc tribometer made by CSM Switzerland, with normal force of 1N, 3N, and 5 N, respectively, a stainless steel ball with a diameter of 6 mm, a dry sliding distance of 100 m, and linear speed of 20cm/s (**Figure 9**). For C-Si-Al composite films, the friction coefficient exhibited values 2–3 times lower than that of the uncoated substrates. An increasing tendency of the friction coefficient with the increase of the applied force was noticed. At 3 N, the film friction coefficient becomes equal with the value for the substrate. Over this value of applied force, film friction coefficient continues to increase.

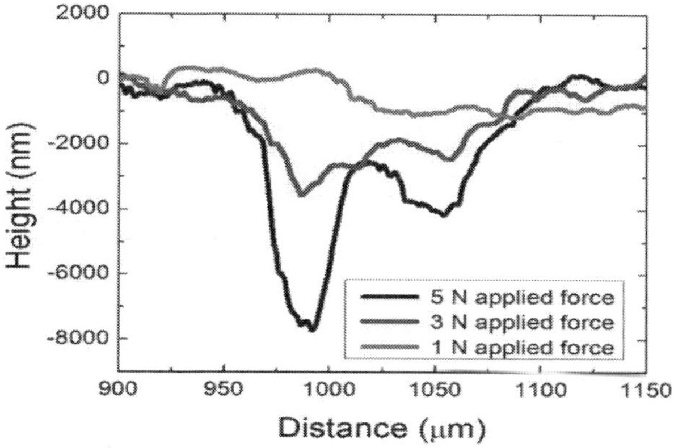

Figure 10. Depth profile of the wearing traces after tribological characterization.

The wear rate was estimated for 3 N and 5 N loading forces based on depth-profile measurements performed on the wearing traces. The wear rate was 1.23E-05 mm^3 /Nm for a force of 3N, while for 5 N it is higher: 6.16E-05 mm^3 /Nm. These values indicate a good behavior of the deposited layers under working conditions. Due to high substrate roughness, the rate for 1N load could not be calculated.

Figure 10 presents the depth profile of the wearing traces after the tribological characterization. The roughness was determined by AFM analysis, with a range of 20–34 nm, confirming along with SEM measurements the nanometric granular structure of the obtained thin film. SEM investigations show a high-quality film surface, without major defects.

6. Surface-free energy properties for binary and ternary composites

Wetting of solid substrates by liquids is a fundamental phenomenon with relevance to both the technological and natural worlds. Applications include the spreading behavior of liquid coatings as well as chemical reactors and thus the understanding and characterization of the solid surfaces wettability is of great interest. The wettability of solid-fluid-vapor interfacial phenomena is determined by the estimation of the contact angle between a solid surface and a liquid drop. This measurement is often used, being a useful and sensitive tool for assessing the wettability of a surface, the solid surface energy, and liquid surface tension [41, 42].

The contact angles and free surface energy were determined by means of Surface Energy Evaluation system (SEE System). The device is based on the usage of a charge-coupled device (CCD) camera, which observes a liquid drop on the studied surface, by using sessile drop method. The tangent angle of the drop with the solid surface is measured by selecting three points on the contour of the drop, two for solid-liquid, and one for liquid-vapor interfaces.

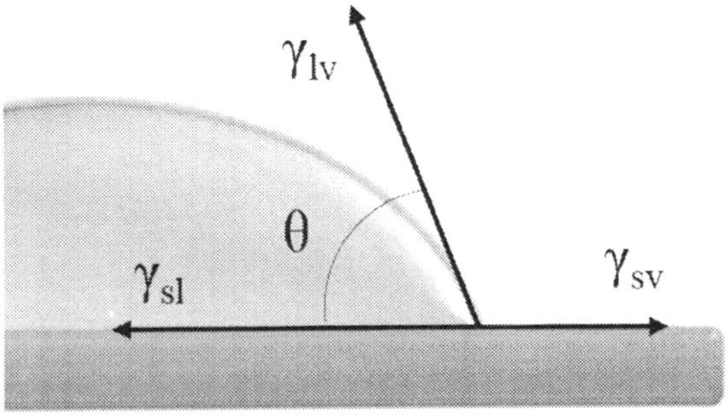

Figure 11. Drop profile and tangent angle of the drop with the solid surface.

This technique is generally applied only to low-energy solid surfaces, as most available liquids as water and ethylene glycol have relatively low surface tensions and wet solids with higher surface energy [43].

In **Figure 11**, the tangent angle of the drop with the solid surface defined as "contact angle" method is presented. It can be interpreted as a mechanical force balance on the three-phase contact line; the surface tension is an energy per unit area, equivalent to a force per unit length acting on the contact line. Horizontal balance of forces is given by Young equation:

$$\gamma_{SV} = \gamma_{SL} + \gamma_{LV} \cos\Theta \tag{1}$$

In order to perform the measurements, the samples were set on a holder in front of the CCD camera, and the static contact angles were measured from the front view. The averaged results were calculated from 10 values and measured on different location on sample, and the drop volume was of 0.4 μl. The constant small volume of the drops used for the measurements is necessary in order to minimize the gravitational effect.

The water contact angles Θ results on the thin films are plotted in **Figure 12**. These results were further analyzed using Owens Wendt method, which allows us to determine the electron-donor and electron-acceptor parameters of the free surface energy. The total free surface energy is a sum of polar (γ^{AB}) and dispersion (γ^{LW}) components:

$$\gamma = \gamma^{LW} + \gamma^{AB} \tag{2}$$

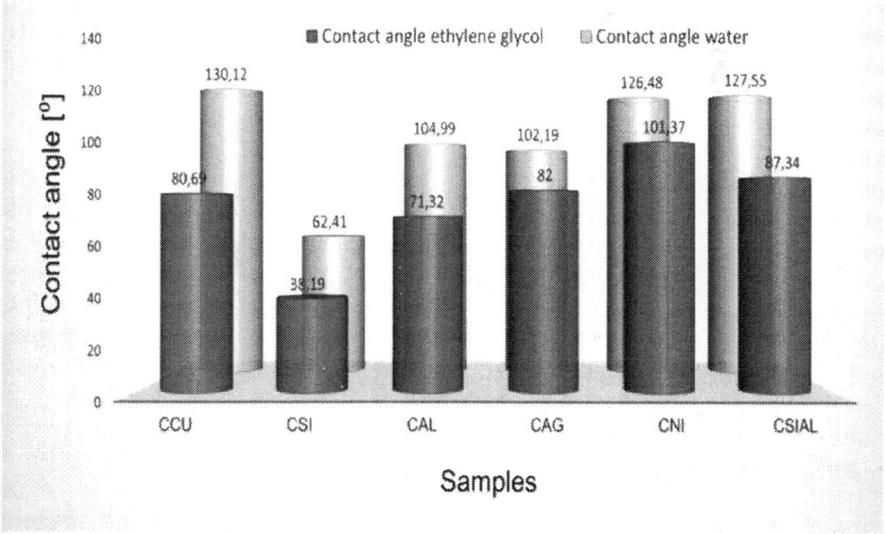

Figure 12. Contact angle measurements.

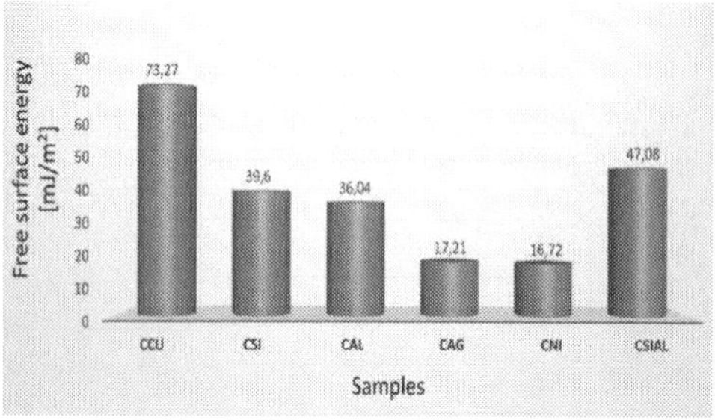

Figure 13. Comparative view of surface-free energies.

where γ^{LW} refers to the total Lifshitz-Van der Walls interaction and γ^{AB} refers to the acid-base interaction [15, 44]. The comparative view of the surface free energy values calculated is presented in **Figure 13**.

The surface free energy can be different from sample to sample due to the nonuniformity of the samples. In order to reduce this, we take instants of the drops for the various liquids used on one sample only, and we calculate the contact angle from these instants.

Figure 14. Images of droplet-stilled water on different thin films.

http://dx.doi.org/10.5772/63367

Typical image frames acquired by the image analysis system described are displayed in **Figure 14**. The figure shows a printout of the side view for a sample during experimental measurement for stilled water on the substrate. These data were digitalized and measured using the software, the omissions of other images obtained, for ethylene glycol, were deliberate to avoid redundancy of images.

Contact angle can reflect the wettability of the materials, and it is influenced by many aspects such as surface characters, roughness, and temperature. Roughness was affected by manufacture craft of the substrate material; however, surface free energy was variable because surface free energy represented the interface molecular between materials and the analyzing liquid. The values of the contact angle reported above indicate a hydrophobic character of the surfaces except for the CSi thin film that has a hydrophilic character.

7. Conclusion

The TVA deposition technique has been successfully used to obtain thin films that presented high purity, low roughness, and good adhesion on the substrates. Also, it is suitable for obtaining carbon-containing binary and ternary composite structures.

The combined deposition of carbon and metal improves basically the film adhesion and the friction coefficient. The structure of C-Me films changes with the metal and its content, as verified by TEM and EDS analysis. HRTEM images and SAED (Selected Area Electron Diffraction) reveal that C-Me structures consist of a distribution of nanocrystallites embedded within an amorphous matrix.

From the tribological analysis, the friction coefficient of the ternary structure showed a 2–3 times decrease with a slight increasing tendency in the range of 0.17–0.58 with an increase in loading force compared with the uncoated substrates. The calculated wear rate was in the range 1.23×10^{-5} and 6.16×10^{-5} mm^3/Nm. The ternary structure hardness was high in the range 3600 to 5000 N/mm^2, decreasing with the depth measurement increase.

Material characteristics presented here enhance the mechanical parts with special features such as higher resistance to wearing and gripping and excellent conformity of the layer with the geometry of the coated parts. These results could have a great impact on the development of advanced materials and many manufacturing industries, as well as expanding the technologically important field of interface science where the control of the film-substrate interface would be critical.

Acknowledgements

This work was supported by a grant of the Romanian National Authority for Scientific Research, CNDI–UEFISCDI, project number 160/2012, PN-II-PT-PCCA-2011-3.2-1453.

Author details

Rodica Vladoiu[1*], Corneliu Porosnicu[2], Aurelia Mandes[1], Ionut Jepu[2], Virginia Dinca[1], Aurelian Marcu[2], Mihail Lungu[2], Gabriel Prodan[1] and Liga Avotina[2]

*Address all correspondence to: rvladoiu@univ-ovidius.ro

1 Ovidius University, Constanţa, Romania

2 National Institute for Lasers, Plasma and Radiation Physics, Bucharest, Magurele, Romania

References

[1] Robertson J: Diamond-like amorphous carbon. Mater. Sci. Eng. R. 2002; 37:129–281. DOI:10.1016/S0927-796X(02)00005-0

[2] Aisenberg S, Chabout R: Ion beam deposition of thin films of diamondlike carbon. J. Appl. Phys. 1971; 42:2953–2958

[3] Bewilogua K, Hofmann D: History of diamond-like carbon films — from first experiments to worldwide applications. Surf. Coat. Technol. 2014; 242:214–225. DOI:10.1016/j.surfcoat.2014.01.031

[4] Bin Masripan NA, Ohara K, Umehara N, Kousaka H, Tokoroyama T, Inami S, Zushi K, Fujita M: Hardness effect of DLC on tribological properties for sliding bearing under boundary lubrication condition in additive-free mineral base oil. Tribol. Int. 2013; 65:265–269. DOI:10.1016/j.triboint.2013.01.016

[5] Lubwama M, Corcoran B, Sayers K, Kirabira JB, Sebbit A, McDonnell KA, Dowling D: Adhesion and composite micro-hardness of DLC and Si-DLC films deposited on nitrile rubber. Surf. Coat. Technol. 2012; 206:4881–4886. DOI:10.1016/j.surfcoat.2012.05.079

[6] Zeng A, Neto VF, Gracio JJ, Qi Hua Fan: Diamond-like carbon (DLC) films as electrochemical electrodes. Diamond Relat. Mater. 2014; 43:12–22. DOI:10.1016/j.diamond.2014.01.003

[7] Polaki SR, Kumar N, Ganesan K, Madapu K, Bahuguna A, Kamruddin M, Dash S, Tyagi AK: Tribological behavior of hydrogenated DLC film: Chemical and physical transformations at nano-scale. Wear. 2015; 338–339:105–113. DOI:10.1016/j.wear.2015.05.013

[8] Xiao Y, Shi W, Han Z, Luo J, Xu L: Residual stress and its effect on failure in a DLC coating on a steel substrate with rough surfaces. Diamond Relat. Mater. 2016; vol 66, p: 23-35 DOI: 10.1016/j.diamond.2016.03.009

[9] Kovács GyJ, Veres M, Koós M, Radnócz G: Raman spectroscopic study of magnetron sputtered carbon–nickel and carbon nitride–nickel composite films: The effect of nickel

on the atomic structure of the C/CNx matrix. Thin Solid Films. 2008; 516:7910–7915. DOI:10.1016/j.tsf.2008.04.081

[10] Laurila T, Rautiainen A, Sintonen S, Jiang H, Kaivosoja E, Koskinen J: Diamond-like carbon (DLC) thin film bioelectrodes: Effect of thermal post-treatments and the use of Ti adhesion layer. Mater. Sci. Eng. C. 2014; 34:446–454. DOI:10.1016/j.msec.2013.09.035

[11] Meškinis Š, Vasiliauskas A, Šlapikas K, Gudaitis R, Andrulevičius M, Čiegis A, Niaura G, Kondrotas R, Tamulevičius S: Bias effects on structure and piezoresistive properties of DLC: Ag thin films. Surf. Coat. Technol. 2014; 255:84–89. DOI:10.1016/j.surfcoat.2014.01.026

[12] Pauleau Y, Thiery F: Deposition and characterization of nanostructured metal/carbon composite films. Surf. Coat. Technol. 2004; 180:313–322.DOI: 10.1016/j.surfcoat.2003.10.077

[13] Wu G, Xu S, Chu P: Tribological behavior of Ti-Al-Si-C-N hard coatings deposited by hybrid arc-enhanced magnetron sputtering. J. Vac. Sci. Technol. A. 2012; 30:021501. DOI: 10.1116/1.3676186

[14] Musa G, Mustata I, Blideran M, Ciupina V, Vlădoiu R, Prodan G, Vasile E, Ehrich H: Thermionic Vacuum Arc (TVA) new technique for high purity carbon thin film deposition. Acta Phys Slovaca. 2005; 55:417–421. WOS: 000230405200009

[15] Vladoiu R, Dinca V, Musa G: Surface energy evaluation of unhydrogenated DLC thin film deposited by thermionic vacuum arc (TVA) method. Eur. Phys. J. D. 2009; 54:433–437. DOI: 10.1140/epjd/e2009-00178-5

[16] Musa G, Mustata I, Ciupina V, Vladoiu R, Prodan G, Vasile E, Ehrich H: Diamond like nanostructured carbon film deposition using thermionic vacuum arc. Diamond Relat. Mater. 2004; 13:1398–1401. DOI: 10.1016/j.diamond.2003.10.048

[17] Surdu Bob C, Vladoiu R, Badulescu M, Musa G: Control over the sp2/sp3 ratio by tunning plasma parameters of the thermionic vacuum arc. Diamond Relat. Mater. 2008; 17:1625–1628. DOI: 10.1016/j.diamond.2008.02.032

[18] Vladoiu R, Ciupina V, Contulov M, Dinca V, Mandes A, Bursikova V: Synthesis and characterization of nanostructured a-C:H thin films by Gaseous Thermionic Vacuum Arc (G-TVA) deposition technique. Plasma Chem. Plasma Process. 2012; 32:219–229. DOI: 10.1007/s11090-011-9344-x

[19] Takeno T, Saito H, Goto M, Fontaine J, Miki H, Belin M, Takagi T, Adachi K: Deposition, structure and tribological behavior of silver–carbon nanocomposite coatings. Diamond Relat. Mater. 2013; 39:20–26. DOI: 10.1016/j.diamond.2013.07.004

[20] Meškinis S, Vasiliauskas A, Šlapikas K, Niaura G, Juškėnas R, Andrulevičius M, Tamulevičius S: Structure of the silver containing diamond like carbon films: Study by multiwavelength Raman spectroscopy and XRD. Diamond Relat. Mater. 2013; 40:32–37. DOI: 10.1016/j.diamond.2013.09.004

[21] Manninen NK, Ribeiro F, Escudeiro A, Polcar T, Carvalho S, Cavaleiro A: Influence of Ag content on mechanical and tribological behavior of DLC coatings. Surf. Coat. Technol. 2013; 232:440–446. DOI: 10.1016/j.surfcoat.2013.05.048

[22] Gayathri S, Kumar N, Krishnan R, Ravindran TR, Dash S, Tyagi AK, Raj B, Sridharan M: Tribological properties of pulsed laser deposited DLC/TM (TM=Cr, Ag, Ti and Ni) multilayers. Tribol. Int. 2012; 53:87–97. DOI: 10.1016/j.triboint.2012.04.015

[23] Gayathri S, Krishnan R, Ravindran TR, Sundari S, Dash S, Tyagi AK, Raj B, Sridharan M: Spectroscopic studies on DLC/TM (Cr, Ag, Ti, Ni) multilayers. Mater. Res. Bull. 2012; 47:843–849. DOI:10.1016/j.materresbull.2011.11.042

[24] Choi HW, Dauskardt R, Lee SC, Lee KR, Oh KH: Characteristic of silver doped DLC films on surface properties and protein adsorption. Diamond Relat. Mater. 2008; 17:252–257. DOI: 10.1016/j.diamond.2007.12.034

[25] Marciano FR, Bonetti LF, Pessoa RS, Marcuzzo JS, Massi M, Santos LV, Trava-Airoldi VJ: The improvement of DLC film lifetime using silver nanoparticles for use on space devices. Diamond Relat. Mater. 2008; 17:1674–1679. DOI:10.1016/j.diamond.2008.03.007

[26] Pardo A, Gómez-Aleixandre C, Celis JP, Buijnsters JG: Friction and wear behavior of plasma assisted chemical vapor deposited nanocomposites made of metal nanoparticles embedded in a hydrogenated amorphous carbon matrix. Surf. Coat. Technol. 2012; 206:3116–3124. DOI: 10.1016/j.surfcoat.2011.12.041

[27] Onoprienko AA, Danylenko MI: Annealing effects in Ag-doped amorphous carbon films deposited by dc magnetron sputtering. Surf. Coat. Technol. 2012; 206:3450–3453. DOI: 10.1016/j.surfcoat.2012.02.004

[28] Yu X, Qin Y, Wang CB, Yang YQ, Ma XC: Effects of nanocrystalline silver incorporation on sliding tribological properties of Ag-containing diamond-like carbon films in multi-ion beam assisted deposition. Vacuum. 2013; 89:82–85. DOI: 10 1016/j vacuum 2011 11 007

[29] Zhang H-S, Endrino JL, Anders A: Comparative surface and nano-tribological characteristics of nanocomposite diamond-like carbon thin films doped by silver. Appl. Surf. Sci. 2008; 255:2551–2556. DOI: 10.1016/j.apsusc.2008.07.193

[30] Wang C, Yu X, Hua M: Microstructure and mechanical properties of Ag-containing diamond-like carbon films in mid-frequency dual-magnetron sputtering. Appl. Surf. Sci. 2009; 256:1431–1435. DOI: 10.1016/j.apsusc.2009.08.103

[31] Hanus J, Drabik M, Hlidek P, Biederman H, Radnoczi G, Slavinska D: Some remarks on Ag/C:H nanocomposite films. Vacuum. 2009; 83:454–456. DOI: 10.1016/j.vacuum.2008.03.101

[32] Zoubos H, Koutsokeras LE, Anagnostopoulos DF, Lidorikis E, Kalogirou SA, Wildes AR, Kelires PC, Patsalas P: Broadband optical absorption of amorphous carbon/Ag

nanocomposite films and its potential for solar harvesting applications. Solar Energy Mater. Solar Cells. 2013; 117:350–356. DOI: 10.1016/j.solmat.2013.06.019

[33] Baba K, Hatada R, Flege S, Ensinger W, Shibata Y, Nakashima J, Sawase T, Morimura T: Preparation and antibacterial properties of Ag-containing diamond-like carbon films prepared by a combination of magnetron sputtering and plasma source ion implantation. Vacuum. 2013; 89:179–184. DOI: 10.1016/j.vacuum.2012.04.015

[34] Jiang B, Tian C, Song G, Pan Q, Wang Z, Shi L, Qiao Y, Fu H: A green route to synthesize novel Ag/C antibacterial agent. Mater. Res. Bull. 2012; 47:458–463. DOI:10.1016/j.materresbull.2011.10.018

[35] Schwarz FP, Hauser-Gerspach I, Waltimo Ts, Stritzker B: Antibacterial properties of silver containing diamond like carbon coatings produced by ion induced polymer densification. Surf. Coat. Technol. 2011; 205:4850–4854. DOI: 10.1016/j.surfcoat.2011.04.078

[36] Lungu CP, Mustata I, Musa G, et al: Unstressed carbon-metal films deposited by thermionic vacuum arc method. J. Optoelectron. Adv. M. 2006; 8:74–77

[37] Ciupina V, Lungu CP, Vladoiu R, et al: Silicon carbide multilayer protective coating on carbon obtained by thermionic vacuum arc method. J. Nanophotonics. 2014; 8:083996. DOI: 10.1117/1.JNP.8.083996

[38] Ciupina V, Vladoiu R, Lungu CP, Dinca V, Contulov M, Mandes A, Popov P, Prodan G: Investigation of the SiC thin films synthesized by Thermionic Vacuum Arc method (TVA). Eur. Phys. J. D. 2012; 66–99. DOI: 10.1140/epjd/e2012-20470-5

[39] Marcu A, Ticoş CM, Grigoriu C, Jepu I, Porosnicu C, Lungu AM, Lungu CP: Simultaneous carbon and tungsten thin film deposition using two thermionic vacuum arcs. Thin Solid Films. 2011; 519:4074–4077. DOI: 10.1016/j.tsf.2011.01.192

[40] Vladoiu R, Ciupina V, Surdu-Bob C, Lungu CP: Properties of the carbon thin films deposited by thermionic vacuum arc. J. Optoelectron. Adv. Mater. 2007; 9:862–866. WOS:000245834800014

[41] Bonn D, Eggers J, Indekeu J: Wetting and spreading. Rev. Modern Phys. 2009; 81:739–805 DOI: 10.1103/RevModPhys.81.739

[42] Vladoiu R, Ciupina V, Dinca V, Musa G: Influence of the operational parameters on the wettability of the DLC films deposited by TVA method. Chem. Listy. 2008; 102:S1463–S1466. ISSN 1213-7103, 0009-2770

[43] Njobuenwu D, Oboho E, Gumus R: Determination of contact angle from contact area of liquid droplet spreading on solid substrate. Leonardo Electronic J. Pract. Technol. 2007; 6:29–38

[44] Navratil Z, Bursikova V, Stahel P, Sira M, Zverina P,On the analysis of surface free energy of DLC coatings deposited in low pressure RF discharge, Czche. J. Phys. 2004; 54:C877–C882. WOS:000226745800009

Chapter 6

Carbon Nanotube (CNT)-Reinforced Metal Matrix Bulk Composites: Manufacturing and Evaluation

Sebastian Suárez, Leander Reinert and
Frank Mücklich

Additional information is available at the end of the chapter

http://dx.doi.org/10.5772/63886

Abstract

This chapter deals with the blending and processing methods of CNT-reinforced metal matrix bulk composites (Al/CNT, Cu/CNT and Ni/CNT) in terms of solid-state processing, referring mainly to the research works of the last ten years in this research field. The main methods are depicted in a brief way, and the pros and cons of each method are discussed. Furthermore, a tabular summary of the research work of the mentioned three systems is given, including the blending methods, sintering methods, the used amount of CNTs and the finally achieved relative density of the composite. Finally, a brief discussion of each system is attached, which deals with the distribution and interaction of the CNTs with the matrix material.

Keywords: Carbon nanotubes, metal matrix composites, reinforcement effect, solid-state processing.

1. Introduction

Composite materials have been in the spotlight of material science and engineering for a long time already. They provide the capability of tailoring their properties by managing very simple variables such as the reinforcement fraction or the processing parameters, among others. Their application area is found in a wide range of dissimilar fields, ranging from bioengineering [1,2] up to aerospace [3]. The applicability relies on the proper selection of both, the matrix phase and the reinforcing phase. By matrix, the phase with the largest volume fraction is meant, whereas the opposite is valid for the reinforcing phase. In the former – according to the specific application – polymers, ceramics or metals are used. Polymer-matrix composites are mainly used in applications where lightweight is required, working in environments that do not present high temperatures. On the opposite, ceramic matrices are used where inertness under high temperatures is required as well as high mechanical properties. Finally, metal matrix composites (MMC) lie in between both application fields, presenting in most cases tailored microstructures (and subsequently, tailored physical properties) and, in certain cases, lightweight.

Reinforcing phases can be of very dissimilar nature. The most widespread phases are usually ceramic fibres and/or particles (i.e. Al_2O_3, TiC), which show very good mechanical properties, thus enhancing the overall mechanical properties of the composite. However, when considering the transfer properties (electrical and thermal), their ceramic nature plays a detrimental role. Furthermore, the fact that the material being subjected to improvement is a metal, with usually very good transport properties, the task becomes indeed non-trivial.

In recent years, the appearance of carbon nanotubes (CNTs) has opened an interesting new field. Since CNTs show intrinsically outstanding physical properties, the aforementioned drawback brought by ceramic reinforcements might be straightforwardly overcome. Yet, the predicted physical properties of CNTs are only realisable if the CNT is in a "perfect" structural state. By "perfect" structural state, it is meant that there are (a) no structural defects on its lattice, (b) no exo- or endohedral contaminants (synthesis/catalysis residues) and (c) the CNTs are in an isolated state (no CNT agglomeration). Those three conditions are quite challenging to achieve in the praxis.

When it comes to CNTs, different synthesis methods have to be considered. CNT synthesis renders unavoidably contaminants (sulphur/amorphous carbon) and catalyst particles. The most common synthesis methods employed are chemical vapour deposition (CVD), arc discharge and laser ablation. The CVD synthesis is the one with the highest capacity to be scaled to industrial quantities and provides a better control on the morphology of the obtained CNTs [4,5] in comparison with other standard synthesis methods. Yet, the defect state is usually high and should be thoroughly analysed before the application.

Most of the recent research works are leaning towards commercially available MWCNTs as starting material for the composite production. Some of the most recurring suppliers are Nanocyl (Belgium), Nanolab Inc. (USA), Iljin Nanotechnology Co. Ltd (Korea), Bayer MaterialScience (Germany), Chengdu Organic Chemicals Co. Ltd (China), Chinananotech Co. Ltd (China) and Hanhwa Nanotech Co. Ltd (Korea) to name just a few [6–31]. But also self-produced CNTs are used for composite manufacturing. For this purpose, catalytic chemical vapour deposition (CCVD) is the most common way to synthesize CNTs that are used as reinforcement phase in composites [6,7,10,32–39]. The MWCNTs generally show a purity between 90% [10] and 99.5% [19], but mostly around 95%. The diameter ranges from 10 nm to 80 nm with a length of 0.5 μm–50 μm [6–39]. All the information about CNT-MMCs in this book chapter is derived from the research works of the last ten years, including some relevant exceptions.

http://dx.doi.org/10.5772/63886

1.1. Blending methods

Considering that commercial CNTs are usually delivered in agglomerated form, different methods have to be employed to disaggregate and blend them with the metallic matrix material. There are a lot of dispersing and blending processes like magnetic stirring [40], nanoscale dispersion processing [18,21,41], colloidal mixing [6,16,22,26–31,41–49], molecular-level mixing [7,9,11,12,14,24,25,50–54], particle composite system mixing [15,55], friction stir processing [56], layer stacking [57], ball milling [8,10,17,19,20,22–24,32–36,38,50,53,58–74], dipping [75] or roller mixing [37,76]. However, three of them have to be pointed out as the most commonly used ways to disperse and blend CNT metal matrix systems. These three processing methods are ball milling, molecular-level mixing and colloidal mixing. The methods will be briefly commented in the following paragraphs.

Ball milling

It is usually performed using a planetary or attrition ball mill. The mixing is done by filling in the Metal powders and reinforcement phase together with some hard balls into mixing jars and rotating the jar with a certain rotational speed (**Figure 1 a**).

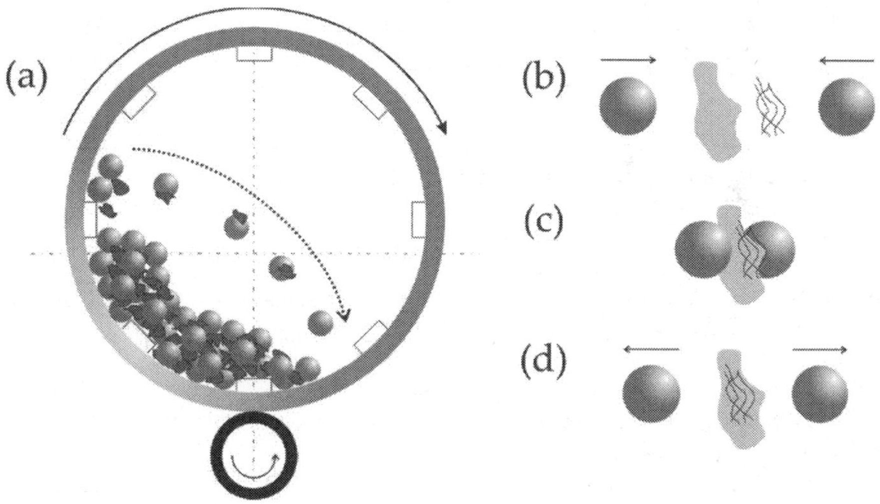

Figure 1. Schematic draft of the ball milling process (a). The used balls are hitting the CNTs and the matrix powder material (b), welding and integrating the two components (c) and bounce of the powder particle to restart this process again at another spot (d).

During the rotational movement, the added balls are falling on top of the powder material (**Figure 1 b**), thus leading to a size reduction, particle welding and integration of CNTs into the matrix powder material (**Figure 1 c** and **d**) generated by the impact energy. Different ball materials and sizes, rotational speeds, balls to powder ratios, gas atmospheres and mixing times can be chosen as main mixing parameters. Usually, a process control agent, for example ethanol, is added in order to prevent cold welding of the matrix material powder particles. In other cases, like mechanical alloying, a cold welding of the particles is desired. Ball milling is known to produce a homogenous distribution of reinforcement phase in metal matrix composites as particle agglomerates are segmented, and dispersed CNTs are partially welded together with the matrix powder material (**Figure 1 d**). In general, high energy ball milling and low energy ball milling are distinguished in literature when it comes to the mixing process of CNTs with metal matrix material. This is because of the main drawback of the method, which is the increasing defect density of CNTs during the mixing process by the direct application of large contact pressures (up to 30 GPa) [77]. Using low energy ball milling, this unwanted effect can be reduced, but it will not vanish. The importance of a low defect density of CNTs will be discussed later on in entry 2.2 [8,10,17,19,20,22–24,32–36,38,50,53,58–74].

Molecular-level mixing

For this mixing method, it is of utmost importance to functionalize the CNTs, for example with an acid treatment (**Figure 2 a**). After this, the CNTs can be dispersed in various solvents, for example using ultrasonic agitation to obtain a stable suspension.

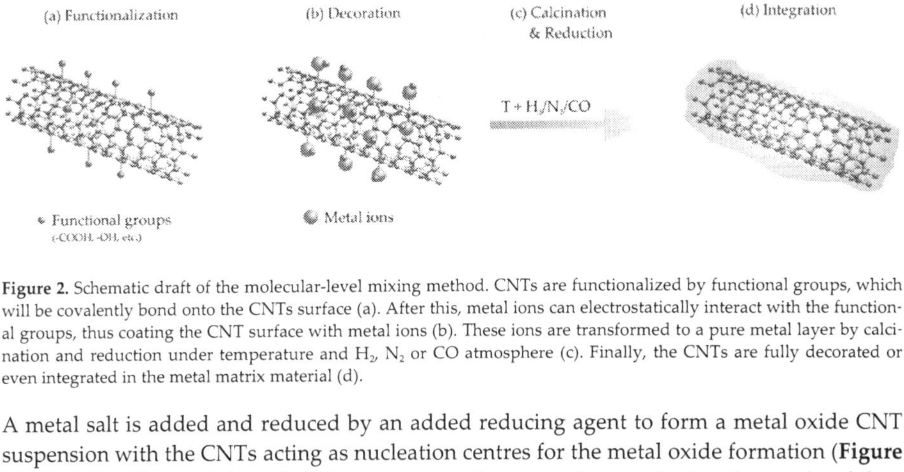

Figure 2. Schematic draft of the molecular-level mixing method. CNTs are functionalized by functional groups, which will be covalently bond onto the CNTs surface (a). After this, metal ions can electrostatically interact with the functional groups, thus coating the CNT surface with metal ions (b). These ions are transformed to a pure metal layer by calcination and reduction under temperature and H_2, N_2 or CO atmosphere (c). Finally, the CNTs are fully decorated or even integrated in the metal matrix material (d).

A metal salt is added and reduced by an added reducing agent to form a metal oxide CNT suspension with the CNTs acting as nucleation centres for the metal oxide formation (**Figure 2 b**). Finally, after washing off all the remaining chemicals, the powder is calcinated and then reduced, for example under hydrogen atmosphere to reduce it to the metal/CNT powder (**Figure 2 c** and **d**). The advantage of this process is that the CNT particles are embedded or coated by the metal matrix, thus resulting in very homogeneous distributions within the metal matrix (**Figure 2 d**). In many other mixing methods, the reinforcement phase can only be situated at the grain boundaries, offering weak interfacial bonding between the reinforcement phase and the matrix material and reducing the homogeneity of the distribution. For the molecular-level mixing process, this is not true, therefore being favoured for applications where a good connectivity of the reinforcement phase network is needed like thermal or electrical applications. The main drawback of the method is the required functionalization, which involves breaking up covalent carbon bonds to add functional groups to the surface of CNTs, thus diminishing their outstanding properties [7,9,11,12,14,24,25,50–54].

Colloidal mixing

For colloidal mixing, the CNTs are dispersed using an ultrasonic bath, homogenizer or magnetic stirrer in a solvent (for the most part, ultrasonic agitation is used (**Figure 3 a**)). There are many different solvents that allow for a fine and stable dispersion of the CNTs, which is discussed in [81] (e.g. DMF or ethylene glycol). The dispersion grade and stability of the CNT suspensions are not only depending on the used solvent, but also on the surface of the used CNTs. Some research works functionalize the surface of the CNTs to obtain electrochemically stable dispersions, which works great but influences the physical properties of CNTs as already discussed for the molecular-level mixing. Other works avoid functionalization, thus using the pristine CNTs, retaining their properties in detriment of the dispersion quality.

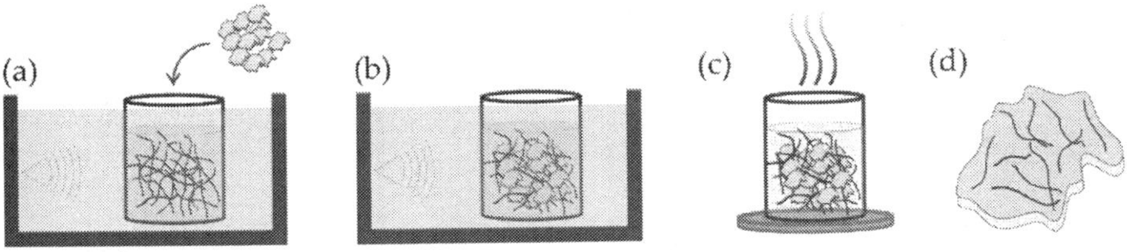

Figure 3. Schematic draft of the colloidal mixing method. First, the CNTs are dispersed in a liquid solvent using ultrasonic agitation, after which the matrix material powder is added (a). The two components are mixed again by ultrasonic agitation (c), and the solvent is evaporated (c) to finally obtain the mixed CNT/metal matrix powder (d).

One point that is controversially discussed is the impact of ultrasonic agitation on the defect density of CNTs. There are studies claiming for a rise in the defect density as a function of the time spent in the ultrasonic bath and others which report the opposite. The observed decrease in the defect density might be a misinterpretation of Raman spectra of disentangled CNTs (which would render improved I_D/I_G ratio stemming from the avoidance of intertube interactions). After the dispersion of CNTs in the solvent, the metal powder is added and mixed with the dispersed CNTs again by ultrasonic agitation, stirring or a homogenizer (**Figure 3 b**). Finally, the solvent is evaporated to obtain a dry mixed powder (**Figure 3 c** and **d**). A significant advantage of this method is that it can be very easily upscaled and still yield the same results [6,16,22,26–31,41–49].

1.2. Processing methods

After merging the CNT/metal composite powders, different densification methods are used for the consolidation of the final samples. Examples of these are cold pressed sintering (CPS) [8,14,17,20,27,28,34,35,47,53,63,74,78], hot uniaxial pressing (HUP) [10,19,26,27,29–31,39,40,47, 49,62,67–70], spark plasma sintering (SPS) [6,7,9,11,12,14–16,18,21,23–25,42,44–46,48,50–52,55,

64,75,79,80], hot or cold rolling [7,50,59,60,64], hot extrusion [18,20,23,40,58,79], high pressure torsion (HPT) [22,31–33,41,54,61,65,66,71,73], friction stir processing [56], hot isostatic pressing (HIP) [22,57], microwave sintering [43], laser engineered net shaping (LENS) [36–38,76] or a combination of those methods. Methods like cold or hot rolling, hot extrusion or HPT are often used for post-processing of the already consolidated samples [7,22,31,50,59,60,64]. The most often used methods for the production of metal matrix composites are CPS, HUP and SPS, which will be described briefly in the following.

Cold Pressed Sintering (CPS)

This is by far the simplest method to densify the blended powder. Using a uniaxial press or an isostatic press, the powder is pre-compacted to the desired shape (**Figure 4**).

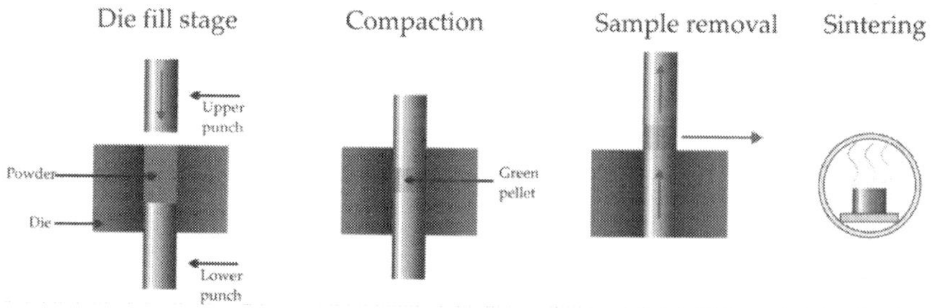

Figure 4. Schematic draft of pressureless sintering or cold pressed sintering. A die is filled with the powder material, which is then pressed uniaxial to a green pellet. The sample is then removed and sintered without pressure in a furnace under vacuum or inert gas atmosphere.

After this, the sample is sintered without further pressure under vacuum or an inert gas atmosphere to form the consolidated sample (**Figure 4**). Even though one heating–cooling cycle might be very time-consuming (the resistive furnaces usually have a very limited heating rate), it has the advantage that many samples of dissimilar shapes can be sintered at the same time, making this sintering method very time efficient. However, the densification mechanism is mainly based on lattice and grain boundary diffusion [81], and large porosities cannot be closed without applying pressure, thus resulting in a poor final density of the composite [8,14,17,20,27,28,34,35,47,53,63,74,78].

Hot Uniaxial Pressing (HUP)

In addition to lattice and grain boundary diffusion (mechanisms for pressureless sintering), plastic deformation and creep can be major sintering mechanisms. As the overall densification rate of a compact is a function of the sum of the active densification mechanisms, pressure-assisted sintering is much more effective than pressureless sintering. The application of an external pressure leads to an increase in the densification driving force and kinetics. As grain growth is not related to the applied external pressure, it is more effective in systems with a

http://dx.doi.org/10.5772/63886

large grain growth to densification rate. To sum up, by using an external pressure, the sintering temperature as well as the sintering time can be reduced [81].

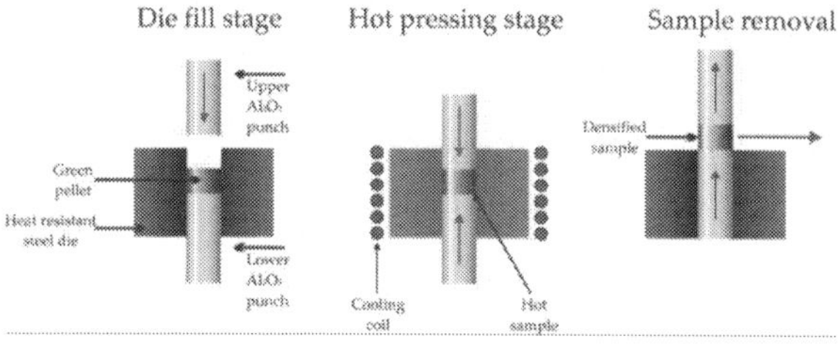

Figure 5. Schematic draft of the hot uniaxial pressing method. A pre-compacted green pellet is inserted in a die, and uniaxial pressure is applied while the pellet is heated by induction under vacuum or inert gas atmosphere. The densified sample is finally removed.

When it comes to HUP, as with the CPS method, the mixed powder is usually pre-compacted using a uniaxial press or an isostatic press to obtain green pellets. After this, the green pellets are typically inserted in a die (often a steel die) where two punches (e.g. alumina) exert a uniaxial pressure onto the sample (**Figure 5**). The heating of the sample is usually conducted by induction, and thus, this system is very limited in the heating rate, rendering HUP as a very time-consuming sintering process. However, high pressures (several hundred MPa) can be applied with this method while sintering and almost full densification can be achieved (as punches with good mechanical properties can be used). To conclude, HUP is a time-consuming way to sinter samples, but it is also very effective when it comes to the final maximum densification of the sample [10,19,26,27,29–31,39,40,47,49,62,67–70].

Spark Plasma Sintering (SPS)

The active sintering mechanisms in SPS do not vary much from the active mechanisms in HUP. Yet, it provides a much quicker way to consolidate the composite powders [81].

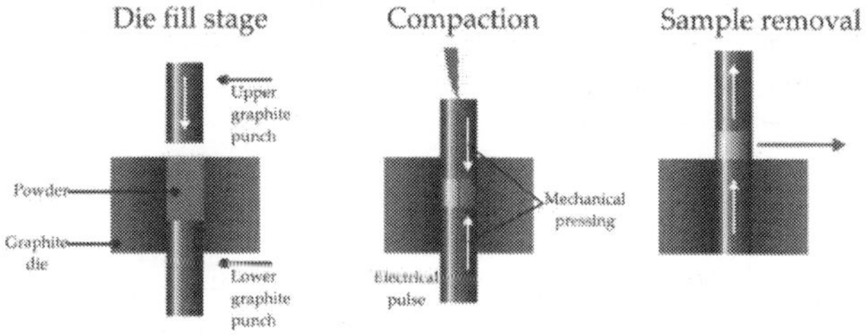

Figure 6. Schematic draft of the Spark plasma sintering method. A graphitic die is filled with the powder material, and a uniaxial pressure is applied via two graphite punches. A pulsed electric DC is applicated, which leads to the heating of the sample by its electrical resistance. The process is conducted under vacuum or inert gas atmosphere.

As with HUP, the mixed powders are pre-compacted using a uniaxial press or an isostatic press to obtain green pellets, which are inserted in a graphite die. With this method, graphite punches are used to allow for inducing a pressure at the sample and conducting a pulsed electric DC through the sample at the same time (**Figure 6**). The sample is heated by its electrical resistance, which depends on the used material. By controlling the used current, the heating rate can be adjusted. This method allows for a very high heating rate of several hundred °C/min, being therefore very time efficient. In contrast to HUP, the used pressure during sintering is much lower (typically about 50 MPa) because of the mechanically weak graphite punches. However, as with HUP, almost full densification can be reached with this method. Overall, this method offers a quick and effective way to consolidate the CNT-reinforced metal matrix powders, therefore being the most employed method in this area [6,7,9,11,12,14–16,18,21,23–25,42,44–46,48,50–52,55,64,75,79,80].

1.3. Potential applications

The main application of the CNT-reinforced metal matrix composites is in structural applications. This is due to the fact that most of the literature is devoted to the study of the mechanical properties of the composites. From the mechanical point of view, the addition of an intrinsically strong second phase would certainly improve the overall properties. Considering that, in the ideal state, the CNTs show a Young's modulus of approximately 1 TPa and a maximum tensile strength of over 60 GPa [82], it is easy to trace their influence. Additionally, a set of factors influences the mechanical behaviour of these composites. First, it has been demonstrated that the addition of CNTs acts on the grain boundary mobility by hindering their displacement during grain growth [83–85]. This effect influences the final microstructure (by refining it) and thus the mechanical behaviour (grain boundary strengthening). Second, a proper distribution of CNTs acts as an obstacle for dislocation movement, activating another strengthening mechanism known as particle dispersion strengthening (Orowan strengthening). Third, it has been shown that the CNTs present a very low or even negative coefficient of thermal expansion (CTE) in a wide temperature range [86,87]. When combined with high CTE materials (such as metals), this CTE mismatch acts also as a strengthening factor. Finally, considering the aforementioned values of the mechanical properties of the CNTs, the strengthening of the composite due to load transfer is expected to be significant. Summarizing, the addition of CNTs to MMCs is clearly expected to be beneficial in terms of the improvement of the mechanical behaviour, and subsequently, it is clear why most of the studies are focused towards this feature. The strengthening effect of the CNTs in MMCs is shown in **Figure 4**. There, the measured yield strength improvement of different systems is given as a function of the CNT volume concentration in the composite. The scattering of results is related to the different mixing and processing methods that were employed. In this context, it becomes clear that after a direct comparison of different research works involving a wide span of production methods for the same system, the task of correlating dissimilar results becomes non-trivial. Due to a lack of available information on the yield strength of Ni-CNT composites, this illustration provides only a summary of the mechanical reinforcement effect that was achieved so far with Al-CNT and Cu-CNT MMC systems.

http://dx.doi.org/10.5772/63886

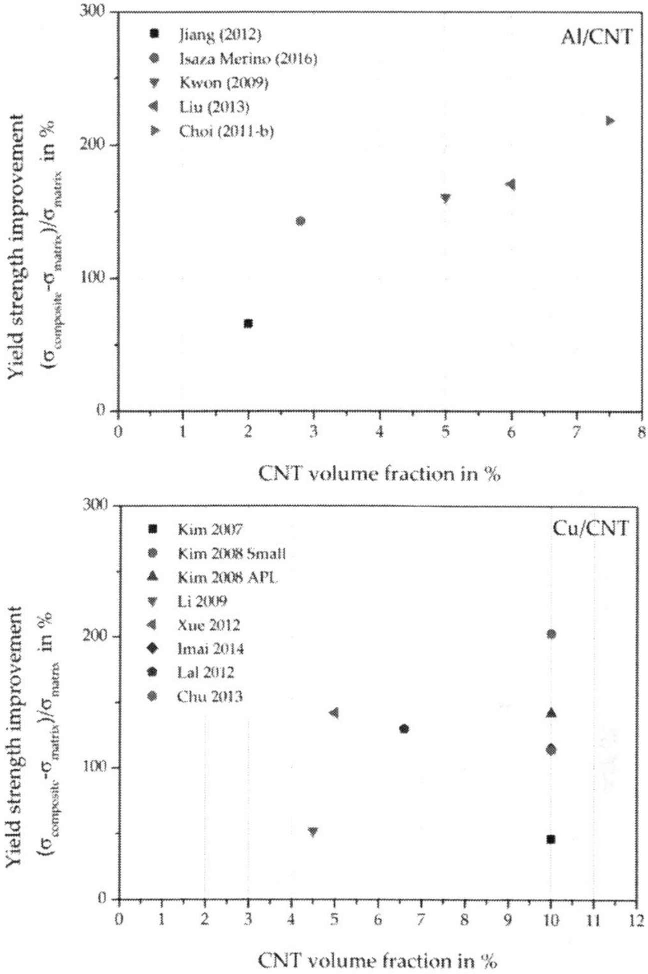

Figure 7. Yield strength improvement against the volume fraction of CNTs in (a) Al/CNT composites [18,20,56,57,60] and (b) Cu/CNT composites [7,9,11,53,64,66,70,75].

However, the transport properties occupy also a large amount of the research in this field. CNTs are predicted to have the highest thermal conductivity known (SWCNT: 6600 W/m.K [88], MWCNT: 3000 W/m.K [89]) and are expected to present a ballistic type of electrical conduction mechanism [90,91]. Both factors are of utmost importance when considering that the material to be improved is a metal (which usually shows very high thermal and electrical properties). However, it is critical to obtain individual CNTs after dispersion, since the transport properties could be reduced up to one order of magnitude If the CNTs are in agglomerated form [92]. In this regard, some parts of the research on Cu-CNT were aimed at the improvement of the thermal properties. Cho et al. observed an improvement of the thermal conductivity for very low CNT concentrations (up to 1 vol. %) [42]. For larger CNT concentrations, the agglomeration of CNTs starts to play a fundamental role in the conductivity decrease. The same effect was observed by Yamanaka et al. for Ni-CNT [46]. Inversely, Firkowska et al. tested several functionalization routes so as to improve the interface in Cu-CNT composites and thus the thermal transport [14]. They observed that for all cases, the degradation of the CNT's intrinsic thermal properties was so relevant and that it was not able to increase the thermal conductivity of the composites in any case.

The electrical behaviour was also studied in several MMC-CNT systems, usually resulting in reduced conductivity. The major factor influencing this behaviour is the presence of agglomerates that, analysed from an electrical transport point of view, are seen as voids. This happens mainly due to the fact that the interfaces between CNT agglomerates and the metal matrices are weak. Only marginal improvements in conductivity were observed for low CNT concentrations in Ni/CNT composites [46].

The tribological properties of CNT-reinforced composites have also been reported in the literature. In most cases, a reduction in both the coefficient of friction (COF) and wear has been observed. In the case of Ni-CNT composites, the COF was reduced in margins from 40 [37] to 75 % [27]. Regarding the Cu-CNT system, the COF reductions ranged from 50 to 75 % [6,8,43, 51,63]. For both systems, the reduction in wear losses achieved in certain cases up to 6 [51], 7 [27] and 8 [43] times that of the reference metal under the same experimental conditions.

The former is usually traced back to a solid lubricant activity of the CNTs during the experiments [37]. Due to their high mechanical properties, they tend to act as rolling second phases at the interface between the rubbing surfaces, thus avoiding the direct contact between the surfaces. In some cases, also the development of a graphitic layer is observed, acting as a solid lubricant [37]. The wear reduction is mainly correlated to the increase in the mechanical properties of the composite, which—as already stated—is generated by a stabilization of the microstructure due to grain boundary pinning. Moreover, Kim et al. stated that a reduction in the wear loss might be due to the reduction in the grain peeling mechanism due to a CNT anchoring of the matrix grains [51].

2. Aluminium/CNT system

2.1. Solid-state processing

Al/CNT composites are mainly interesting, because of their high potential being a lightweight, reinforced material, which can be used in manifold applications. Therefore, Al/CNT composites have been in the focus of research since 1998 [40] and the interest is still growing.

There is a large variety of starting materials on the market that have been used to fabricate the composites. Some of the most mentionable suppliers for Al powders are ECKA Granules Japan Co. Ltd (Japan), Aluminium Powder Company Ltd (United Kingdom), Alpha Industries

(South Korea) and AlfaAesar (Germany) [18,19,21,58,61,62]. The range of used powders goes from several μm up to 75 μm in mean particle size, having different particle shapes and various purity grades between >99% and >99.99% [18,22,39–41,50,56–59,79]. The most used blending method for Al/CNT composites is ball milling [19,20,22,23,50,58–62].

A detailed overview of the research papers in the Al/CNT composite manufacturing can be found in **Table 1**.

Reference	Blending method	Sintering method	CNT content *Value* wt%	Relative density *Value* %
Kuzumaki et al. (1998) [40]	**Stirring:** Stirring in ethanol at 300 rpm for 0.5h, drying in vacuum furnace.	**HUP and hot extrusion:** Packed in an Al case and preheated for 1.5 h at 873 K in vacuum (0.53 Pa) and then compressed with 100 MPa in steel dies for 60 min. The heating and loading rates were 29.1 K/min and 10 MPa/min, respectively. Then, the composites are extruded at 773 K (extrusion ratio = 25 : 1) at a speed of 10 mm/min	5 vol% 10 vol%	94% 96.2%
Xu et al. (1999) [39]	**Hand grinding:** For > 30 min.	**HUP:** At 793 K under a pressure of 25 MPa for more than 30 min.	1 wt% 4 wt% 10 wt%	XXX
Tokunaga et al. (2008) [41]	**Colloidal mixing process:** Mixing by sonication for 5min and then evaporation of the solvent.	**HPT:** At room temperature with an applied pressure of 2.5GPa. The rotation is initiated 5s after the load application and terminated after 30 turns.	5 wt%	98.4%
Kwon et al. (2009)18	**NSD:** Consisting of commercial gas atomized Al powder, CNTs and natural rubber. The precursor is heat treated at 500 °C for 2 h in an argon atmosphere (1 l/min) to evaporate the natural rubber.	**SPS and hot extrusion:** Sintering at 600 °C, holding time of 20 min, heating rate of 40 °C/min and pressure of 50 MPa. The sintered compact was extruded in a 60° conical die at 400 °C with a pressure of 500 kN. The extrusion velocity and the extrusion ratio were fixed at 2mm/min and 20, respectively.	5 vol%	Sintered: 96.1% Hot extruded: 98%
Esawi et al. (2009) [58]	**Ball milling:** At 200 rpm for 3h and 6h using 75 stainless steel milling balls (10mm diameter); giving a ball-to-powder weight	**Hot extrusion:** Compacted at 475MPa. Hot extrusion of the compact is conducted at 500°C using an extrusion ratio of 4:1.	2 wt%	XXX

Reference	Blending method	Sintering method	CNT content *Value* wt%	Relative density *Value* %
	ratio of 10:1. The jars are filled with argon and 2ml of methanol is added.			
Kwon et al. (2010) [79]	**NSD:** Consisting of commercial gas atomized Al powder, MWCNTs and natural rubber. The precursor was heat treated at 500 °C for 2 h in an argon atmosphere (1 l/min) to evaporate the natural rubber.	**SPS and hot extrusion:** Sintered at 480, 500, 560, and 600°C with a heating rate and holding time of 40 °C/min and 20 min, respectively. A pressure of 50 MPa is used. SPSed compacts were extruded in a 60° conical die at 400 °C with a 500 kN press. The extrusion velocity and extrusion ratio were fixed at 2mm/min and 20, respectively.	1 vol%	Up to 96.8%
Choi et al. (2011) [59]	**Ball milling:** A2024 chips are ball-milled at 500 rpm under argon atmosphere for up to 48 h. The ball-to-chip weight ratio was 15:1. A control agent of 1.0 wt% stearic acid is added. The composite powder is then produced by ball-milling 18-h A2024 powder and CNTs	**Hot rolling:** Heated to a temperature of 450 °C and then hot rolled with every 12% reduction per a pass; the initial thickness was 20mm, and the final thickness was 1 mm.	1 vol% 2 vol% 3 vol%	XXX
Choi et al. (2011) [60]	**Ball milling:** Aluminium powder was solely ball-milled for 18h or 12h and then mixed with CNTs by ball-milling for 6h. Or Aluminium powder is directly mixed with CNTs by ball milling for 6h.	**Hot rolling:** Heated to a temperature of 450°C, 480°C or 530°C and then hot rolled with every 12% reduction per a pass; the initial thickness was 20mm and the final thickness was 1 mm.	1.5 vol% 3 vol% 4.5 vol% 6 vol% 7.5 vol% 9 vol%	96.3%-99.9% depending on parame ters.
Kwon et al. (2011) [19]	**Ball milling:** For 3h under an argon atmosphere; 360 rpm, Ø 10mm ball, 10:1 ball to powder weight ratio, and 20 wt% heptane was used as the process control agent.	**HUP:** Consolidating at 500 °C for 5min under a uniaxial pressure of 57 MPa.	5 vol% 10 vol% 15 vol%	100%
Jiang et al. (2012) [20]	**Ball milling:** Flake powder is obtained by ball milling of near-spherical powder with an initial ball-to-powder weight ratio	**CPS and Hot extrusion:** Compacted under 500 MPa pressure, and then sintered in flowing Ar atmosphere at 550 °C for 2h.	0.5 vol% 2 vol%	After CPS: 80-85% After

Reference	Blending method	Sintering method	CNT content *Value* wt%	Relative density *Value* %
	of 20:1 and 423 rpm for several hours in flowing Ar atmosphere with water cooling. The as-prepared Al nanoflakes are first surface modified by PVA with 1700–1800 repeat units and then mixed with a CNT suspension, which is then heated in flowing Ar atmosphere at 500 °C for 2 h to remove the PVA. 1kg blend in 8 hours.	Then heated to 440 °C with a heating rate of 10 °C/min within a vacuum furnace installed with the extruder. And extruded with an extrusion ratio of 20:1 at a ram speed of 0.5 mm/min.		hot extrusion: > 99.5%
Nam et al. (2012) [50]	**Molecular-level mixing and ball milling:** Poly-vinyl alcohol aqueous solution and CNTs are mixed by tumbler ball milling for 48 h, followed by drying in vacuum at 100 °C. The PVA-coated CNTs, $Cu(CH_3COO)\bullet 2\ H_2O$ and 2 M NaOH aqueous solution are dispersed in water and heated to 80 °C to form a CNT/CuO composite. This is followed by vacuum filtering and reduction at 300 °C under a hydrogen atmosphere. CNT/Cu composite powders and Al powders are then mixed using a planetary mill for 3 h with rotation speed of 200 rpm; the ball to powder ratio was 10:1. The composition of the Al–Cu matrix is Al-4wt%.	**SPS and cold rolling:** Sintered at 500 °C for 5 min in a vacuum of 10^{-3} torr under a pressure of 50 MPa . The CNT/Al–Cu composites are solution heat treated at 550 °C for 12 h and quenched in a water bath at room temperature. They were then cold rolled to achieve 5% plastic deformation. The composites are finally aged at 130 °C for 0 to 24 h.	2 vol% 4 vol%	XXX
Kwon et al. (2013) [21]	**NSD:** The precursor was heat-treated at 500°C for 2 h in an argon atmosphere (1 l/min) in order to evaporate the natural rubber.	**SPS:** Sintering at 600 °C, holding time 20 min, heating rate 40 °C/min, and pressure 50 MPa. Then heat-treated in an alumina pan at 670 °C and at 800 °C for 1 h in an argon atmosphere (1 l/min) inside a tube furnace.	5 vol%	96%
Liu	**Friction stir processing (FSP):**	**Friction stir processing:** A	1.6 vol%	XXX

Reference	Blending method	Sintering method	CNT content *Value* wt%	Relative density *Value* %
et al. (2013) [56]	Plates of aluminium alloy are used as the substrate materials. Holes with a depth of 3.5 mm and a diameter of 0 mm, 2 mm, 4 mm, 6 mm, 8 mm and 10 mm respectively were drilled in the aluminium plates, which are then filled with various quantities of CNTs.	tilt angle of 2° is applied on the fixed pin tool during FSP. The pitch distance was 0.3 mm. The forward velocity of the rotating pin is kept as a constant of about 30 mm/min, and a rotational speed of 950 rpm is used. Five passes on the same position were conducted on each plate.	2.5 vol% 4.4 vol% 5.3 vol% 6 vol%	
Asgharzadeh et al. (2014) [61]	**Ball milling**	**HPT:** Pre-compacted under pressure of ~1 GPa. Then, the green compacts are compressed and deformed under the applied pressure of 6 GPa at room temperature. The rotation of the lower anvil was started after 30 seconds of load application with a rotational speed of 1 rpm and was terminated after various numbers of revolutions up to 15.	3 vol%	98.5% - 99.5%
Phuong et al. (2014) [22]	**Colloidal mixing and Ball milling:** CNTs are functionalized and dispersed in ethanol by sonication. Then, Al powders and CNT suspension were mixed by mechanical stirring and simultaneously heated in order to evaporate most of the ethanol. The slurry is then processed in a high energy ball mill for 2 h with 250 rpm. Drying is accomplished at 60 °C using a vacuum oven.	**HIP and HPT:** After cold compaction, capsule free method HIP is accomplished under pressure of 103 MPa, at 600 °C for 30 min. Straining was carried out at room temperature, by torsion to 5 revolutions, under pressure of about 5 GPa and at rotation speed of 2 rpm. The specimens are then annealed in vacuum for 3 h at temperatures of 100 °C, 150 °C, 200 °C and 250 °C.	0.5 vol% 1 vol% 1.5 vol% 2 vol%	XXX
Isaza Merino	**Layer stacking:** CNTs are added to a solution	**HIP:** In argon atmosphere . The pressure and temperature are	0.5 wt% 2 wt%	XXX

Reference	Blending method	Sintering method	CNT content *Value* wt%	Relative density *Value* %
et al. (2016) [57]	composed of 4 wt.%, PVA dissolved in distilled water. This mixture is magnetically stirred at 1000 rpm during 1 h, followed by a sonication at energies between 40 and 60 kJ and a power of 100W. Finally, this mixture is cured at room temperature in a Petri dish for 8 days. The composite sheets are cut into small sections and stretched at a rate of 2mm/min in a tensile machine at 80 °C. Then, two composite sheets of polymer/CNTs are alternately stacked with three aluminium sheets.	raised gradually during 1.5 h to 40MPa and 650 °C for 30 min. This allows for the evaporation of PVA and aluminium diffusion between the sheets to finally produce the consolidation of the composite.		
Carvalho et al. (2016) [62]	**Ball milling:** Al-Si 88-12 wt% and CNTs are mixed using 12 steel milling balls with 10-mm diameter, presenting a ball-to-powder weight ratio of 10:1. The jar was placed in a rotation device, and the mixing was made with a constant rotation speed of 40 rpm, during 6 days (low-energy ball milling).	**HUP:** Uniaxial load pressure of 35MPa and a temperature of 550°C during 10min.	2 wt% 4 wt% 6 wt%	XXX
Chen et al. (2016) [23]	**Ball milling:** Al powder is milled with 2 wt.% stearic acid. The powder is sealed in a ZrO_2 jar together with 600 g ZrO_2 media balls (10mm in diameter) in an argon gas atmosphere. The rotation speed is 200 for 240 min. Then, Al-flake powder	**SPS and hot extrusion:** Sintering at 800 K, 850 K and 900 K. The heating rate is 20 K/min and the holding time at target temperatures is 60min. A pressure of 30MPa is applied on the sample under a vacuum of ~5 Pa. Before	1 wt%	>99%

Reference	Blending method	Sintering method	CNT content *Value* wt%	Relative density *Value* %
	is mechanically bathed in isopropyl alcohol-based solution with ~1 wt.% zwitterionic surfactants and 1 wt.% CNTs in a plastic bottle on a rocking ball milling machine for 120 min. Thirty-gram ZrO_2 media balls were added to assist the coating of CNTs on flaky powders.	hot extrusion, the sintered sample is preheated to 700 K and kept for 180 s under an argon gas atmosphere. Then, the billet is immediately put into a steel container) and extruded through a die. The extrusion ratio and the ram speed are 37:1 and 3mm/s, respectively.		

Table 1. Summary of blending and sintering methods for the production of Al/CNT composites.

2.2. Distribution and interaction with the matrix material

In all the analysed articles, CNT agglomeration is observed, being the size and amount of those agglomerates strongly related to the chosen dispersion method. Specifically, in those reports where ball milling is used, a significant clustering is noticed. This might be due to the use of non-functionalized CNTs that tend to easily re-agglomerate during processing. Yet, the utilization of functionalized CNTs in ball milling mixed blends would not bring any further improvement, since the structural quality of the CNTs would be significantly lower. However, despite the presence of unavoidable agglomerates, very fine cluster dispersion and a homogeneous distribution can be observed. Interestingly, one would assume that the formation of clusters would be detrimental to the proper densification of the composites. However, as already depicted in Table 1, the final densities of the composites were never below 94%. The formation of clusters is one of the most challenging problems to overcome when dealing with CNTs, since a trade-off between an optimal initial agglomerate disentanglement and the lowest possible damage to the CNT structural state must be found. There are two ways to observe the amount of agglomerates in the composites. Certain reports focus on the evaluation by electron microscopy on the polished surface of the samples, whereas other authors evaluate fracture surfaces of the composites. Both approaches are valid, in the sense that they show a good overview of the agglomeration. However, SEM imaging of the surface would allow a further quantification of the agglomerates size and distribution by segmenting the C-containing phases. Bakshi proposed an interesting methodology to quantify the dispersion, based on distance calculation algorithms from electron micrographs, which can be found in **Bakshi et al.** [93].

If we focus our attention on the CNT degradation during blending and after densification, we observe dissimilar results obtained by Raman spectroscopy, even for the same manufacturing method. For example, in samples densified by SPS (a technique known to exert strong thermomechanical stress on the CNTs), negligible variation of the I_D/I_G ratio is reported [23,79]

http://dx.doi.org/10.5772/63886

as opposed to an increase in the defect density in other reports [18]. It is therefore very difficult to discern whether the sintering process has indeed any sort of influence on the CNT degradation. It has been already reported that given the strong stresses from SPS, CNTs are usually either damaged or degraded within the matrix material [94]. When strong shear forces are applied as in the case of HPT, sharp increases in the D band intensity are measured, depicting a modification of the graphitic structure. In all cases, an increased amount of structural damage would favour the reactivity of the CNT shells.

The CNT degradation is a major issue when working with metals which are strong carbide builders [95]. It is well known that aluminium tends to form a stable carbide (Al_4C_3) in a wide compositional range [96], being of brittle nature and having the major drawback of being water soluble [97]. Some authors mention that the formation of this carbide would not be as detrimental as expected [18,58], since it only sacrifices the outermost CNT layer, providing an optimal interface with the matrix. However, despite being able to efficiently transfer the applied load, the utilization of the composite in structural applications in humid environments would lead to an interfacial degradation. Another way to obtain this carbide is by transitioning from Al_2O_3 to Al_4C_3 [98]. Yet, this phase transition occurs only under certain circumstances that are rarely achieved in solid-state processing.

3. Copper/CNT system

3.1. Solid-state processing

Reports on the solid-state processing of Cu/CNT systems deal with a variety of different raw materials, blending methods and sintering techniques, therefore sticking out in the research field of CNT-reinforced metal matrix composites.

When it comes to the used Cu starting material, a great many of commercially available powders or chemicals are employed, making it very hard to compare the Cu/CNT systems of different publications. The materials used depend directly on the blending and production process that has been used. The range goes from nanosized to almost 100 μm sized Cu powder particles, having different particle shapes (spherical, dendritic) and various purity grades between >99.5% and >99.95% [6–12,14,15,32–35,42–44,50,52–55,63–71]. The most used blending methods are ball milling, molecular-level mixing or colloidal mixing, each having their advantages and disadvantages as discussed in the introduction [6–12,14–17,32–35,42–45,51–55,63–75]. But especially for Cu, molecular-level mixing is employed very often, using copper compounds in solvents that have to be chemically treated in order to become pure copper (e.g. copper(II) acetate monohydrate, Cu(II) sulphate pentahydrate or simply CuO). Some prominent suppliers for the Cu starting materials are Sigma-Aldrich (USA), Alfa Aesar GmbH & Co. KG (Germany), Chang Sung Co. (Korea), Junsei Chemical Co. Ltd (Japan), Kojundo Chemical Lab. Co. Ltd (Japan), TLS Technik GmbH (Germany) or New Materials Research Co. Ltd (China) [6,12,14–16,32,33,42,54].

Therefore, to review the progress in Cu/CNT composites, a detailed comparison of blending methods, sintering techniques and achieved final relative composite densities has to be conducted, which can be found in **Table 2**.

Reference	Blending method	Sintering method	CNT content	Relative density
Tu et al. (2001) [63]	**Ball milling:** CNTs are milled for 8 h in an organic liquid. Then, CNTs are sensitized, activated and finally coated by electroless nickel. Copper- and nickel-coated CNTs are mixed for 30 min in a ball mill.	**CPS:** Isostatically pressed at a pressure of 600 MPa at 100 °C for 10 min under vacuum. Then sintered at 800 °C for 2 h.	4 vol% 8 vol% 12 vol% 16 vol%	97.5% 97.5% 97% 95%
Chen et al. (2003) [35]	**Ball milling:** CNTs are sensitized, activated and then coated by electroless nickel for 15 min. Then, copper powder and nickel-coated CNTs are mixed by ball milling for 30 min.	**CPS:** Isostatically pressed at a pressure of 600 MPa at 100 °C for 10 min under vacuum. Then sintered at 800 °C for 2 h.	4 vol% 8 vol% 12 vol% 16 vol%	97.5% 97.5% 97% 95%
Kim et al. (2006) [64]	**Ball milling:** Cu powder and CNTs are mixed through high energy ball milling process for 24h with 150rpm.	**SPS and cold rolling:** Pre-compacted under a pressure of 10 MPa and then sintered at 700 °C for 1 min in vacuum (0.13 Pa) under a pressure of 50 MPa with a heating rate of 100 °C/min. Finally, the samples are cold rolled up to 50% reduction, followed by full annealing at 650 °C for 3 h.	5 vol% 10 vol%	99.3% 99.1%
Kim et al. (2007) [51]	**Molecular-level mixing process:** CNTs are purified and functionalized by acid treatment. Then, Cu acetate monohydrate is added and sonicated for 2h. The solution is vaporized with magnetic	**SPS:** Sintered at 550°C for 1 min in vacuum(0.1 Pa) with a pressure of 50MPa and a heating rate of 100 K/min.	5 vol% 10 vol%	99% 99%

Reference	Blending method	Sintering method	CNT content	Relative density
	stirring at 100 °C, followed by calcination at 350 °C in air. and reduction under N_2 atmosphere.			
Kim et al. (2008) [9]	**Molecular-level mixing process:** CNTs were purified under sonication for 10 hours and then functionalized. Then, 20 mg of CNTs are sonicated in 300ml ethanol for 2 hours. $Cu(CH_3COO)\bullet 2\ H_2O$ is added and sonicated again for 2 hours. After drying at 250 °C in air and calcinating at 300 °C in air, the powder is reduced under a hydrogen atmosphere.	**SPS:** Pre-compacted under a pressure of 10 MPa and then sintered at 550°C for 1 minute in vacuum (0.13 Pa) with a pressure of 50 MPa. The heating rate is 100°C/ min.	5 vol% 10 vol%	98 ± 2% 98 ± 1.5%
Kim et al. (2008) [11]	**Molecular-level mixing process:** CNTs are cleaned and functionalized. Then, 10.5 mg of CNTs are dispersed in 100 ml of oleylamine in an ultrasonic bath for 3 h. 3 g of copper acetate monohydrate is added and the solution is purged with argon for 1 h. Then, the mixture is heated to 523 K for 10 min with a heating rate of 10 K/min. After cooling, the powders are reduced at 573 K for 2 h under hydrogen atmosphere.	**SPS:** Sintering at 823 K for 1 min in vacuum (0.13 Pa) with a pressure of 50 MPa and a heating rate of 100 K/min.	5 vol%	XXX
Daoush (2008) [12]	**Molecular-level mixing process:** Copper (II) sulphate pentahydrate, tri-sodium citrate monohydrate and CNTs are stirred with 500 rpm magnetic stirrer	**SPS:** Sintered at 600°C for 1 min with a pressure of 20 MPa under vacuum (0.13 Pa).	10 vol% 20 vol% 30 vol% 40 vol%	97.9% 97% 96. 7% 96%

Reference	Blending method	Sintering method	CNT content	Relative density
	for 2 h at room temperature to suspend the CNTs in the solution. The equivalent formaldehyde amount is added with a rate of 0.1 ml/min within 30 min. The solution is filtered, washed and vacuum dried at 100°C for 2 h and finally reduced at 400°C for 30 min under hydrogen atmosphere.			
Chai et al. (2008) [15]	**Particle Composite System mixing:** CNTs are cleaned in nitric acid in the ultrasonic bath. Mixing condition is 5000 rpm in rotary speed for 40 min.	**SPS:** Sintered at 600 °C for 5 min. The heating rate is 100 °C/min and a pressure of 50 MPa is applied.	5 vol% 10 vol% 15 vol%	98.4% 98.6% 96.1%
Li et al. (2009) [65]	**Ball milling:** In an argon protected environment at room temperature. The mass ratio of ball to powder is 10:1, and the milling time is 5 h.	**HPT:** Pre-compacted with 500 MPa in air at room temperature. Then consolidated under 6 GPa for 5 revolutions at room temperature.	1 wt%	XXX
Li et al. (2009) [66]	**Ball milling:** In an argon protected environment for 5 h	**HPT:** Pre-compacted and then consolidated under 6 GPa for 5 revolutions at room temperature.	1 wt%	XXX
Chu et al. (2010) [55]	**Particle Composite System mixing:** CNTs are purified in nitric acid in an ultrasonic bath, then filtered,washed and dried at 120°C. The mixing is performed with 5,000 rpm for 40 min.	**SPS:** Sintered at 550°C-650°C in vacuum (<5 Pa) with a pressure of 40-60 MPa for 5-10 min. The heating rate is 100°C/ min.	5 vol% 10 vol%	96.2% - 99.2% 95.3% - 98.9%
Cho et al. (2010) [42]	**Colloidal mixing process:** CNTs are sonicated in acid solution at 323 K for 24h. Then, Cu powder and CNTs were sonicated separately in ethanol solution for 1 h,	**SPS:** Sintered at 823 K for 1 min under a uniaxial pressure of 50 MPa. The heating rate is 50 K/min.	0.5 vol% 1 vol% 1.5 vol% 2 vol% 3 vol% 5 Vol%	96.8 - 99.0%

Reference	Blending method	Sintering method	CNT content	Relative density
	blended, stirred for 30 min and oven-dried at 323 K. The powders are finally heat-treated at 623 K for 1 h in Ar–5% H2 atmosphere.		10 vol%	
Uddin et al. (2010) [67]	**Ball milling:** Milled in Ar atmosphere and for different hours with 50 stainless steel balls (diameter of each ball: 10 mm, ball to powder ratio 10:1, milling speed: 200 rpm). For bronze composites, the chemical composition is Cu 79%, Sn 10%, Zn 3% and Ni 8%.	**HUP:** Sintering at 750 °C with pressure of 40 MPa for Cu–CNT composites in Ar-atmosphere, whereas 800 °C and 40 MPa for Bronze–CNT composites.	Cu 0.1 wt% Cu 0.5 wt% Cu 1 wt% Cu 2 wt% Cu 4 wt% Bronze 0.1 wt% Bronze 0.5 wt% Bronze 1 wt% Bronze 2 wt% Bronze 4 wt%	96.8% 96.8% 93.7% 91% 82.6% 95.9% 93.8% 88.9% 89.9% 85.6%
Xu et al. (2011) [8]	**Ball milling:** Milled for 5 h in an organic liquid.	**CPS:** Cold pressed at 200 MPa, then sintered at 850 °C in vacuum atmosphere for 5 h. After cooling, a second pressing at 600 MPa and a second sintering are performed.	10 vol%	96%
Jenei et al. (2011) [33]	**Ball milling**	**HPT:** Pre-compacted by cold pressing. Then consoli dated at RT and 373 K with a pressure of 2.5 GPa and 10 revolutions.	3 vol%	97%
Kim et al. (2011) [52]	**Molecular-level mixing process:** CNTs are cleaned and functionalized. Then, the CNTs are dispersed within ethanol by sonication. Cu $(CH_3$- COO)• $2H_2O$ is added to	**SPS:** Sintered at 823 K f **CPS:** Pre-compacted using a hydraulic press 130 kN. Then s intered in a tube furnace at 600 °C in vacuum (10.6 Pa) or 1 min in vacuum (0.13 Pa) with a	5 vol% 10 vol%	XXX

Reference	Blending method	Sintering method	CNT content	Relative density
	the CNT suspension and sonicated for 2 h. After vaporization at 333–373 K and calcination at 573 K in air, the powders are reduced at 523 K for 4 h under hydrogen atmosphere.	pressure of 50 MPa.		
Rajkumar et al. (2011) [43]	**Colloidal mixing process:** CNTs are purified and oxidized using nitric acid treatment by sonication for 10 min at 60°C Then, the CNTs are sensitized and activated. The activated CNTs are introduced into the electroless copper bath and stirred for 30 min while using sonication. The coated CNTs are dispersed in ethanol with vigorous sonication for 10min after which the copper powder is added. The solution is stirred and evaporated at 120°C. The powder is finally mixed in an electric agate pestle mortar for 2 h.	**Microwave Sintering:** Pre-compacted in a hydraulic press. Sintering at 800°C with the soaking time of 5 min with a ramp rate of 12 °C/min.	5 vol% 10 vol% 15 vol% 20 vol%	95.5% 96% 96.5% 94%
Guiderdoni et al. (2011) [44]	**Colloidal mixing process:** CNTs are dispersed in deionized water using a sonotrode for a few seconds. The Cu powder is added. After further sonication of 1 min, the dispersion is immersed in liquid N_2 for 2 min and freeze-dried at- 40 °C for 48 h in vacuum (12 Pa).	**SPS:** Sintered in vacuum (<10 Pa) at 700°C with a heating rate of 100 °C/min for 6 min. The pressure of 100 MPa is gradually applied within the first minute of the dwell and maintained during the remaining 5 min. Samples are naturally cooled.	0.5 vol% 1 vol% 2 vol% 3 vol% 4 vol% 5 vol% 10 vol% 16 vol%	95% 96% 98% 96% 93% 93% 88% 78%
Pham et al. (2011) [34]	**Ball milling:** The CNTs are treated in acid at 60 °C for 4 h. Then, the CNTs are dispersed in acetone. The Cu powder and the CNTs dispersed in acetone are mixed through a	**CPS:** Sintering at temperatures of 850, 900 and 950° C for 2 h in argon atmosphere.	0.5 wt% 1 wt% 1.5 wt% 2 wt% 2.5 wt% 3 wt%	XXX

Reference	Blending method	Sintering method	CNT content	Relative density
	high-energy ball milling process for 6 h at 300 rpm.		3.5 wt%	
Firkowska et al. (2011) [14]	**Molecular-level mixing process:** CNTs are functionalized by polymer wrapping and dispersed in 1 wt% Poly(sodium 4-styrenesulfonate) by sonication for 2 h and stirring overnight. Then, CNTs are oxidized, washed and suspended in water by ultrasonic treatment. Copper acetate is added. After 12 h, the suspension is heated up to 70 °C to evaporate the solvent. After calcination, the mixture is finally reduced under H_2 at 350 °C for 2 h	**CPS:** Pre-compacted using a hydraulic press 130 kN. Then sintered in a tube furnace at 600 °C in vacuum (10.6 Pa) for 60 minutes. **SPS:** Pre-compacted under a pressure of 6 kN. Then sintered at 600 °C for 5 min in a vacuum a pressure of 50 MPa and a heating rate of 100 °C/min.	0.2 wt% 1 wt% 3 wt% 10 wt%	XXX
Xue et al. (2012) [7]	**Molecular-level mixing process:** Dispersion of CNT in ethanol (1 g/l) by sonication for 30 min. $Cu(CH_3COO) \bullet 2H_2O$ (2g/ml) is dissolved in $NH_3 \bullet H_2O$ (40%), mixed and sonicated for 30 min with the CNT dispersion. After solvent vaporization and calcination, the powder is reduced at 250 °C for 2 h under hydrogen atmosphere.	**SPS and hot rolling:** At 550 °C (vacuum 0.1 Pa) under a pressure of 50 MPa for 5 min with a heating rate of 100 °C/min. Then hot rolled up to 50% reduction at 650 °C.	5 vol%	XXX
Shukla et al. (2012) [10]	**Ball milling:** Using stainless steel balls of Ø 3/8 inch (ball to powder ratio 5:1) under argon with 200 rpm for 20h.	**HUP:** Sintering at 700 °C with 30 MPa pressure under vacuum (0.001 Pa) for 30 minutes.	5 vol%	93.4%
Chu et al. (2013) [68]	**Ball milling**: Mixed with Cu–Ti powders with 1200 rpm for 2h. Ball-to-powder weight ratio is 10:1. Alcohol is added.	**HUP:** At 760 °C for 20 min under an uniaxial pressure of 40 MPa	5 vol% 10 vol% 15 vol%	99% 98% 96%
Guiderdoni et al. (2013)	**Colloidal mixing process:** CNTs are dispersed in deionized	**SPS:** At 700°C (vacuum <10 Pa) under a pressure of 100	5 vol% 8,4 vol%	96% 85% 82% 73%

Reference	Blending method	Sintering method	CNT content	Relative density
[6]	water using a sonotrode (a few s econds). Cu powder is then added and sonicated for one minute. The dispersion is immersed in liquid N_2 for 2 min and freeze-dried at -40 °C for 48 h in vacuum (12 Pa).	MPa for 6 min with a heating rate of 100 °C/min. Then natural cooling.	17,1 vol% 33,2 vol%	
Jenei et al. (2013) [32]	**Ball milling**	**HPT:** Pre-compacted by cold pressing and consolidated at RT and 373 K with 2.5 GPa and 10 revolutions.	3 vol%	97%
Lal et al. (2013) [53]	**Molecular-level mixing process and Ball milling:** CNTs are functionalized in acidic solution. After washing and drying, they are dispersed in Sodium dodecyl sulphate (SDS). And coated with copper. Copper powder is ball milled for 10 h at 500 rpm using stainless steel balls (5 mm). Ball to powder ratio is 10:1 and ethanol is taken as a process control reagent. Copper-coated CNTs are milled together with previously milled Cu powder for 5 h at 250 rpm.	**CPS:** Pre-compacting with 500 MPa pressure. Then sintering at 900°C for 4 h under vacuum (1.33 Pa).	0.5 wt% 1 wt% 1.5 wt%	XXX
Tsai et al. (2013) [54]	**Molecular-level mixing process:** CNTs are functionalized and treated with 250 ml of oleylamine at room temperature for 2.5 h. 3 g of copper (II) acetate monohydrate are added and the mixture is sealed and purged in an argon-protected environment for 4 h. After heating to 523 K with 10K/min for 15 min, the powder is cooled, calcinated at 573 K in air for 2 h and then reduced in a N_2 atmosphere.	**HPT:** Rotation speed of 1 rpm with a load of 2.5 GPa at room temperature for 5 revolutions. The rotation is initiated 5 s after the load application.	5 vol%	XXX

Reference	Blending method	Sintering method	CNT content	Relative density
Shukla et al. (2013) [69]	**Ball milling:** Using stainless steel balls of 10 mm diameter (ball to powder ratio 5:1) under argon atmosphere at 200 rpm for 20 h.	**HUP:** Sintering at 700°C for 30 min using 30 MPa uniaxial stress under vacuum (0.001 Pa). The compacts produced underwent delamination in the centre on cooling and were repressed at 725°C under same conditions.	0.2 vol% SWCNT 5 vol% SWCNT 10 vol% S WCNT 5 vol% MWCNT 10 vol% MWCNT	95% 99% 97% 96% 98%
Chu et al. (2013) [70]	**Ball milling:** Mixed with Cu–0.76wt.% Cr alloying powder under argon atmosphere with a rotary speed of 1200 rpm for 120 min. A ball-to-powder weight ratio of 10:1 is used, and alcohol is added.	**HUP:** Pre-compacted to a green density of 75% and then consolidated at 750 °C for 15 min with a heating rate of 50 °C/min and a pressure of 40 MPa .	5 vol% 10 vol% 15 vol%	XXX
Yoon et al. (2013) [71]	**Ball milling:** Using stainless steel balls for 1 h (ball-to-powder weight ratio was 15:1) and 15 rpm.	**HPT:** Under room temperature and 6 GPa using 10 revolutions at a constant speed of 1 rpm.	5 vol% 10 vol%	99% 99%
Sule et al. (2014) [45]	**Colloidal mixing process:** Copper powder and CNTs are dispersed separately in 60 ml ethanol using sonication. CNTs were sonicated for 1 h, stirred for 5 min, sonicated again for 30 min and stirred again for 3 min. Copper powder is sonicated in ethanol for 1 h and stirred for 2 min. The slurries are mixed together, sonicated for 1 h and stirred for 3 min. After drying using a rotary evaporator, the powder is annealed for 30 min at 550 °C with a heating rate of 5 °C/min under inert Ar atmosphere to reduce the oxygen content.	**SPS:** Sintered at 650°C under a vacuum of 1 hPa. A constant pulse-to-pause ratio of 10:5 (10 ms on and 5 ms off) is applied under a pressure of 50 MPa.	1 vol%	97%
Barzegar Vishlaghi	**Ball milling:** CNTs are sonicated in acetone for 20	XXX	XXX	XXX

Reference	Blending method	Sintering method	CNT content	Relative density
(2014) [72]	and cleaned in nitric acid for 1 h. Cu and Fe powder is milled for 15h. Then, CNTs are added and milled a gain for 15h using chromium steel balls under argon atmosphere with 300 rpm and a ball to powder weight ratio of 20:1.			
Imai et al. (2014) [75]	**Dipping process:** Dipping Cu0.5 Ti powder into a zwitterionic surfactant water solution (3-(N,N-dimethyl stearylammonio) propanesulfonate) with CNTs. The powders are then dried at 353 K for 7200 s in air. To remove surfactant films, powders are heated to 873K in H_2 and Ar atmosphere.	**SPS:** Sintered at 1223 K for 1800 s under 30 MPa in vacuum (6 Pa). The compacts are preheated at 1073K for 800s with a heating rate of 1K/s in Ar atmosphere. After preheating, immediately extruded by using hydraulic press under an extrusion ratio of 12.8.	0.19 wt% 0.34 wt%	XXX
Akbarpour et al. (2015) [73]	**Ball milling:** In Ar atmosphere for 3 h at 200 rpm using 10 mm balls, with a 10:1 ball to powder weight ratio and 0.5 wt% of stearic acid.	**HPT:** Pre-consolidated at a load of 150 MPa and a temperature of 973 K with a holding time of 30 min. Then, HPT process at room temperature with 6 GPa and various numbers of revolutions.	4 vol%	99.6%
Chen et al. (2015) [17]	**Ball milling:** The CNT and $NbSe_2$ are coated with copper by electroless plating. CNTs and $NbSe_2$ are mixed with Cu powder mechanically.	**CPS:** Powders are cold pressed with 200 MPa and sintered at 750 °C for 2 h in N_2 shielding. The sintered compact is crushed, repressed at 400 MPa and sintered again at 750 °C for 2 h in N_2 shielding.	1-4 vol% CNT with various amounts of copper-coated $NbSe_2$	XXX
Sule et al. (2015) [16]	**Colloidal mixing process:** CNTs are sonicated in ethanol for 1 h and stirred for 5 min using a magnetic stirrer. Then, Ru powder is added, combined with further sonication for 1 h	**SPS:** Sintering at 600 or 650°C with a heating rate of 80°C/min, a holding time of 5 min and a pressure of 50 MPa.	1 vol% CNT (600°C) 2 vol% CNT (600°C)	98.28% 98.15% 97.72% 96.77% 96.15% 97.63%

Reference	Blending method	Sintering method	CNT content	Relative density
	and stirring for 3 min. Copper powder was ultrasonicated in ethanol for 1 h and stirred for 2 min. The slurries are mixed together and are sonicated for 1 h and stirred for 3 min. After drying, the powders are further blended in a Turbula T2F mixer for 1h at a speed of 101 revolutions per minute and finally annealed for 30 min at 550 °C with a heating rate of 5 °C/min under inert Ar atmosphere to reduce the oxygen content.		0.5 vol% CNT 0.5 vol% Ru (600°C) 1 vol% CNT 0.5 vol% Ru (600°C) 2 vol% CNT 0.5 vol% Ru (600°C) 1 vol% CNT 0.5 vol% Ru (650°C) 2 vol% CNT 0.5 vol% Ru (650°C)	97.08%
Varo et al. (2015) [74]	**Ball milling:** Flake Cu–CNT nanocomposite powders are synthesized by 5h of ball milling, using a rotation speed of 300rpm; the ball-to-powder ratio is 5:1.	**CPS:** Pre-compacted with 200 MPa of uniaxial pressure at room temperature. Then, the green pellet is warm-pressed at 500 MPa for 1 h at 500°C and finally sintered in a tube furnace at 950°C for 2 h under argon atmosphere.	0.5 wt% 1 wt% 0.15 wt% 0.2 wt% 0.25 wt% 3 wt% 5 wt%	Decreasing density with increasing CNT content.

Table 2. Summary of blending and sintering methods for the production of Cu/CNT composites.

3.2. Distribution and interaction with the matrix material

The only stable copper carbide known is the copper acetylide Cu_2C_2 [99]. It is usually observed as a transition phase during the purification of Cu and after the reaction of cuprous oxide with water [99]. In the studied Cu-CNT composites, no carbide was detected whatsoever. However,

Reference	Blending method	Sintering method	CNT content	Relative density
	out for 24 h.			
Scharf et al. (2009) [37]	**Roller mixing:** Mixed in a twin-roller mixer consisting of two rolls rotating in opposite directions (one clockwise and one anticlockwise). This mixing is carried out for 24 h.	**LENS:** Pulsed Nd:Yttrium aluminium garnet laser (power rating of 500 W, wavelength of 1.064 nm) is focused on the substrate to create a melt pool into which the powder feedstock is delivered through Ar gas (mass flow rate 3.5 l/min). Laser power of 400 W, 35 A current, hatch width of 6.35 mm, layer thickness of 0.01 in.	10 vol%	XXX
Hwang et al. (2009) [36]	**Ball milling:** For 48 hours using two different sizes of tungsten carbide balls of diameter 12.7 and 6.35 mm.	**LENS:** Pulsed Nd:YAG laser (power rating of 500 W, wavelength of 1.064 nm) is focused on the substrate to create a melt pool into which the powder feedstock is delivered through Ar gas. Laser power of 300 W, 30 A current.	5 wt%	XXX
Singh et al. (2010) [38]	**Ball milling:** For 48 hours using two different sizes of tungsten carbide balls of diameter 12.7 and 6.35 mm.	**LENS:** Pulsed Nd:YAG laser (power rating of 500 W, wavelength of 1.064 nm) is focused on the substrate to create a melt pool into which the powder feedstock is delivered through Ar gas.	15 vol%	XXX
Suárez et al. (2012) [47]	**Colloidal mixing process:** Merging of the dispersion of the CNTs in an ultrasound bath with N,N-dimethylfor mamide (DMF) and the subsequent addition of Ni powder. The CNT concentration is 0.023 mg/ml. After 10min in the ultrasound bath, the resulting solution was stirred to homogenize the mixture and then dried in a ventilated furnace. The powder is finally	**CPS and HUP:** Uniaxially pressed with a pressure of 990 MPa. Then sintered in a vacuum (2 Pa) tube furnace at 950°C for 2.5 h. Or sintered in a hot uniaxial press in vacuum (0.2 Pa) at 750°C for 2.5h with an axial pressure of 264MPa.	1 wt%	93.7% (CPS) 95.5% (HUP)

Reference	Blending method	Sintering method	CNT content	Relative density
	milled in an agate mortar.			
Hwang et al. (2013) [25]	**Molecular-level mixing process:** CNTs were purified and functionalized by using HF, acid solution mixture of H_2SO_4/HNO_3. After cleaning and drying, they are dispersed in ethylene glycol solution by ultra-sonication and $Ni(C_2H_3O_{2)2}$•$4H_2O$ is added. Then 2 M-NaOH and hydrazine are injected into the mixed solution, heated to 70 °C and kept 30 min. Finally, the reduction process of fabricated CNT/Ni composite powders is performed in H_2 + CO mixed atmosphere for 2 h at 400°C.	**SPS:** Pre-compacted under a pressure of 10 MPa. Sintered at 700 °C for 1 min under vacuum (0.13 Pa) employing a pressure of 50 MPa. The heating rate is maintained at 100 °C/ min.	6 vol%	98-99%
Suárez et al. (2013) [78]	XXX	**CPS:** Manufactured by cold pressing at 990 MPa and sintering at 950 °C for 2.5 h in vacuum. .	1 wt% 3 wt% 5 wt%	95-96%
Suárez et al. (2013) [26]	**.Colloidal mixing process:** Merging of the dispersion of the CNTs in an ultrasound bath with ethylene glycol and the subsequent addition of Ni powder. The CNT concentration is 0.2 mg/ml. After 10min in the ultrasound bath, the resulting solution was stirred to homogenize the mixture and then dried in a ventilated furnace. The powder is finally milled in an agate mortar.	**HUP:** Cold pressed with 990 MPa. Then sintered in a hot uniaxial press under vacuum (0.2 Pa) at 750 °C for 2.5 h with a 264MPa axial pressure.	1 wt% 2 wt% 3 wt% 5 wt%	95-97%
Sairam et al. (2014) [80]	No mixing, only pure Ni (high purity 99.99%) with a size range from 1 to 5 μm.	**SPS:** Pre-compacted under a pressure of 5 MPa. Then sintered at 700, 850 or 1000 °C for 5 min under a controlled argon atmosphere under a	XXX	96.8-97.7% (700°C) 97.8-99% (850°C)

Reference	Blending method	Sintering method	CNT content	Relative density
		pressure of 50, 65 or 80 MPa. The heating rate is maintained at 100°C/min.		98.1 -99.2% (1000°C)
Borkar et al. (2014) [24]	**Ball milling or molecular-level mixing:** High energy ball milling for 24 h at 400 rpm. For the MLM process, CNTs are purified and functionalized by acid treatment. After cleaning and drying, they are dispersed in ethylene glycol by ultra-sonication and $Ni(C_2H_3O_{2})_2 \bullet 4H_2O$ is added. Then, 2 M-NaOH and hydrazine are injected, the solution is heated to 70 °C and kept 30 min. Finally, the CNT/Ni composite powders are reduced in H_2 + CO mixed atmosphere for 2 h at 400°C.	**SPS:** Pre-compacted under a pressure of 5 MPa. Then sintered at 800 °C for 5 min under controlled argon atmosphere under a pressure of 80 MPa. The heating rate is maintained at 100° C/min.	5 vol%	XXX
Nguyen et al. (2014) [48]	**Colloidal mixing process:** Functionalization of the metal powders and CNTs by zwitter-ionic surfactant, 3-(N,N-dimethylstearylammonio)-propane-sulfonate (ZW). Suspensions with varying concentration of ZW (0.04-0.10 wt%) in de-ionized water are heated to 353 K for 10 min with a fixed CNT concentration (0.1 wt%) and ultrasonicated for 15 min. The metal powder is submersed in 100 ml of the 0.06 wt% ZW-0.1 wt% CNT suspensions with varying amounts of dwell time from 1 to 6 h. The ZW-CNT-metal powder suspensions are then dried at 423 K in vacuum, and the dried powders are subsequently annealed in a vacuum	**SPS:** Consolidated at 550 or 650 °C in a vacuum environment of 6 Pa, using a holding time of 3 min and a pressure of 200 MPa.	0.1 wt%	93% (550°C) >97% (650°C)

Reference	Blending method	Sintering method	CNT content	Relative density
	furnace at 693 K for 10 min to decompose a majority of the surfactant.			
Suárez et al. (2014) [27]	**.Colloidal mixing process:** Merging of the dispersion of the CNTs in an ultrasound bath with ethylene glycol and the subsequent addition of Ni powder. The CNT concentration is 0.2 mg/ml. After 10min in the ultrasound bath, the resulting solution was stirred to homogenize the mixture and then dried in a ventilated furnace. The powder is finally milled in an agate mortar.	**CPS or HUP:** Consolidated with an axial pressure of 990 MPa. Then densified by pressureless sintering (CPS) under vacuum (1 Pa) at 1223 K for 2.5 h. The second set is densified in a hot uniaxial press (HUP) under vacuum (0.1 Pa) at 1023 K also for 2.5 h.	1 wt% 2 wt% 3 wt% 5 wt%	95.5% (CPS) 98% (HUP)
Suárez et al. (2014) [28]	**.Colloidal mixing process:** Merging of the dispersion of the CNTs in an ultrasound bath with ethylene glycol and the subsequent addition of Ni powder. The CNT concentration is 0.2 mg/ml. After 10 min in the ultrasound bath, the resulting solution was stirred to homogenize the mixture and then dried in a ventilated furnace. The powder is finally milled in an agate mortar.	**CPS:** Cold pressed at 990 MPa. Then sintered at 950 °C for 2.5 h in vacuum.	1 wt% (6.5 vol%) 2 wt% (12 vol%) 3 wt% (18 vol%) 5 wt% (27 vol%)	XXX
Suárez et al. (2014) [29]	**.Colloidal mixing process:** Merging of the dispersion of the CNTs in an ultrasound bath with ethylene glycol and the subsequent addition of Ni powder. The CNT concentration is 0.2 mg/ml. After 10min in the ultrasound bath, the resulting solution was stirred to homogenize the mixture and then dried in a ventilated furnace. The powder	**HUP:** Cold pressed with 990 MPa and then sintered in under vacuum (0.2 Pa) for 2.5 h at 750°C with a 264 MPa axial pressure.	1 wt% (6.5 vol%) 2 wt% (12 vol%) 3 wt% (18 vol%) 5 wt% (27 vol%)	Up to 99%

Reference	Blending method	Sintering method	CNT content	Relative density
	is finally milled in an agate mortar.			
Rossi et al. (2014) [49]	**.Colloidal mixing process:** Merging of the dispersion of the CNTs in an ultrasound bath with ethylene glycol and the subsequent addition of Ni powder. The CNT concentration is 0.2 mg/ml. After 10min in the ultrasound bath, the resulting solution was stirred to homogenize the mixture and then dried in a ventilated furnace. The powder is finally milled in an agate mortar.	**HUP:** Cold pressed with 990 MPa and then sintered in under vacuum (0.2 Pa) for 2.5 h at 750°C with a 264 MPa axial pressure.	1 wt% (6.5 vol%) 2 wt% (12 vol%) 3 wt% (18 vol%)	XXX
Suárez et al. (2015) [31]	**Colloidal mixing process:** Dispersion of CNTs in ethylene glycol (0.2 mg/ml) is mixed in an ultrasound bath with the metallic powders for 20 min and then evaporated to obtain a metal–CNT blend.	**HUP and HPT:** Sintered in vacuum (0.1 Pa) for 2.5 h at 750°C at a compacting pressure of 264 MPa. The densified composites were further processed at room temperature inducing severe plastic deformation by HPT with 4 rotations applying a 4 GPa pressure. For the thermal stability studies, the HPT-processed samples are annealed at 573 K with a heating rate of 20°C/min and a dwell time of 3 h.	1 wt% 3 wt%	XXX
Reinert et al. (2015) [30]	**Colloidal mixing process:** CNT concentration in ethylene glycol is 0.006 vol%. A homogenizer (5-min treatment) combined with an ultrasonic bath (20-min treatment) is used for the dispersion of the CNTs. After that, Ni powder is added and mixed with the CNT	**HUP:** Pre-compacted at 990 MPa. Then sintered in a hot uniaxial press (axial pressure of 264 MPa) in vacuum (0.2 Pa) at 750°C for 2.5 h.	6.5 vol%	98%

Reference	Blending method	Sintering method	CNT content	Relative density
	dispersion for 5 min using the homogenizer. Finally, the solvent is evaporated at 150°C in a furnace.			

Table 3. Summary of blending and sintering methods for the production of Ni/CNT composites.

4.2. Distribution and interaction with the matrix material

As observed in the other metal-CNT systems, agglomeration is observed in all cases. If the different blending methods are considered, it is very unlikely to obtain a predominantly individual CNT dispersion rather than a homogeneous cluster distribution. Nevertheless, this cluster distribution provides also diverse improvements with regard to microstructural refinement and properties enhancement. Furthermore, when the initial agglomerate size (as-received state) is considered, all dispersion methods render very good disaggregation. Regarding the damage to the CNTs during processing, it is clear that highly energetic processing as ball milling increases the amount of defects on the CNTs, whereas milder processing routes as colloidal mixing present a fairly unmodified state of the CNTs structure. Interestingly, the sintering process tends to help in the defect healing [29]. The application of temperature in a non-reactive environment (vacuum sintering of non-carbide forming metal matrices) diminishes the I_D/I_G ratio as observed with Raman spectroscopy. Furthermore, the purity index $I_{G'}/I_D$ also is reduced, evidencing some sort of contaminant removal (e.g. amorphous carbon).

Nickel does not form stable carbides [101]; however, it has been reported that under certain conditions it is possible to obtain Ni_3C [102]. This carbide is brittle and, despite improving the interfacial cohesion to the CNTs, would have a detrimental influence on the transport properties of the composite. Due to its metastable nature, it is quite complex to detect it in the composite. It has been shown that the most suitable way is to assess the crystallographic lattice of the Ni in the vicinity of the CNT-containing areas by selected area electron diffraction [29,102]. In the case of a C-Ni reaction, the originally FCC Ni phase would transition to an intermediate Ni_3C (hcp) and later would stabilize in an hcp Ni phase. Thus, if this hcp lattice is detected in the near region of the CNT zones, it would mean that the Ni_3C was previously present. To the date, Ni_3C in Ni-CNT systems has been only detected once by Hwang et al. [102] under a very specific set of synthesis conditions.

Yamanaka et al. and Nguyen et al. show that the application of SPS rendered a very smooth interface between the CNTs and the matrix [46,48]. This was also observed in HR-TEM for hot pressed samples [29]. In general, all the different approaches have resulted in seamless interfaces that are later translated in improved mechanical properties [26,48], thermal properties [46], tribological [27,37] and thermal expansion behaviour [47,78]. Additionally, a proper interface would favour a grain boundary drag that improves the microstructural control by achieving refined microstructures [26,28,48].

5. Outlook

Although a large amount of efforts was directed towards the development of CNT-MMCs systems, there is still a significant room for improvement. This statement is supported by the fact that as described, dissimilar results have been obtained by even using the same processing methods as well as the same type and amount of CNTs. This is generated by a scarcity of proper knowledge of each particular system. As an example, there is still an ongoing discussion in the community about the most suitable dispersion and blending methods for a certain application. Furthermore, it can be noticed that the impact of interphases on the physical properties of the composites is still not well understood. Thus, we foresee a very high potential to gain new insights on each particular system and subsequently achieve further developments in this field.

Acknowledgements

The present work is supported by funding from the Deutsche Forschungsgemeinschaft (DFG, projects: MU 959/38-1 and SU 911/1-1). The authors wish to acknowledge the EFRE Funds of the European Commission for support of activities. This work was also supported by the CREATe-Network Project, Horizon 2020 of the European Commission (RISE Project No. 644013).

Author details

Sebastian Suárez*, Leander Reinert and Frank Mücklich

*Address all correspondence to: s.suarez@mx.uni-saarland.de

Functional Materials, Dept. of Materials Science and Engineering, Saarland University, Germany.

References

[1] Ramakrishna S, Mayer J, Wintermantel E, et al. Biomedical applications of polymer-composite materials: A review. *Compos Sci Technol* 2001; 61: 1189–1224.

[2] Scholz MS, Blanchfield JP, Bloom LD, et al. The use of composite materials in modern orthopaedic medicine and prosthetic devices: A review. *Compos Sci Technol* 2011; 71: 1791–1803.

[3] Cavalier JC, Berdoyes I, Bouillon E. Composites in Aerospace Industry. *Adv Sci Technol* 2006; 50: 153–162.

[4] Jorio A, Dresselhaus M, Dresselhaus G. *Carbon Nanotubes: Advanced topics in the synthesis, structure, properties and applications*. Heidelberg, DE: Springer-Verlag, 2008.

[5] Prasek J, Drbohlavova J, Chomoucka J, et al. Methods for carbon nanotubes synthesis —review. *J Mater Chem* 2011; 21: 15872.

[6] Guiderdoni C, Pavlenko E, Turq V, et al. The preparation of carbon nanotube (CNT)/ copper composites and the effect of the number of CNT walls on their hardness, friction and wear properties. *Carbon* 2013; 58: 185–197.

[7] Xue ZW, Wang LD, Zhao PT, et al. Microstructures and tensile behavior of carbon nanotubes reinforced Cu matrix composites with molecular-level dispersion. *Mater & Des* 2012; 34: 298–301.

[8] Xu W, Hu R, Li J, et al. Effect of electrical current on tribological property of Cu matrix composite reinforced by carbon nanotubes. *Trans Nonferrous Met Soc China* 2011; 21: 2237–2241.

[9] Kim KT, Cha S Il, Gemming T, et al. The role of interfacial oxygen atoms in the enhanced mechanical properties of carbon-nanotube-reinforced metal matrix nanocomposites. *Small* 2008; 4: 1936–1940.

[10] Shukla AK, Nayan N, Murty SVSN, et al. On the Possibility of Occurrence of Anisotropy in Processing of Cu-CNT Composites by Powder Metallurgical Techniques. *Mater Sci Forum* 2012; 710: 285–290.

[11] Kim KT, Eckert J, Menzel SB, et al. Grain refinement assisted strengthening of carbon nanotube reinforced copper matrix nanocomposites. *Appl Phys Lett* 2008; 92: 121901.

[12] Daoush WM. Processing and characterization of CNT/Cu nanocomposites by powder technology. *Powder Metall Met Ceram* 2009; 47: 531–537.

[13] Dai P-Q, Xu W-C, Huang Q-Y. Mechanical properties and microstructure of nanocrystalline nickel-carbon nanotube composites produced by electrode position. *Mater Sci Eng A* 2008; 483-484: 172–174.

[14] Firkowska I, Boden A, Vogt A-M, et al. Effect of carbon nanotube surface modification on thermal properties of copper-CNT composites. *J Mater Chem* 2011; 21: 17541–17546.

[15] Chai G, Sun Y, Sun J 'Jenny', et al. Mechanical properties of carbon nanotube–copper nanocomposites. *J Micromechanics Microengineering* 2008; 18: 035013.

[16] Sule R, Olubambi P A., Sigalas I, et al. Spark plasma sintering of sub-micron copper reinforced with ruthenium–carbon nanotube composites for thermal management applications. *Synth Met* 2015; 202: 123–132.

[17] Chen B, Yang J, Zhang Q, et al. Tribological properties of copper-based composites with copper coated NbSe2 and CNT. *Mater Des* 2015; 75: 24–31.

[18] Kwon H, Estili M, Takagi K, et al. Combination of hot extrusion and spark plasma sintering for producing carbon nanotube reinforced aluminum matrix composites. *Carbon* 2009; 47: 570–577.

[19] Kwon H, Bradbury CR, Leparoux M. Fabrication of Functionally Graded Carbon Nanotube-Reinforced Aluminum Matrix Composite. *Adv Eng Mater* 2011; 13: 325–329.

[20] Jiang L, Li Z, Fan G, et al. The use of flake powder metallurgy to produce carbon nanotube (CNT)/aluminum composites with a homogenous CNT distribution. *Carbon* 2012; 50: 1993–1998.

[21] Kwon H, Takamichi M, Kawasaki A, et al. Investigation of the interfacial phases formed between carbon nanotubes and aluminum in a bulk material. *Mater Chem Phys* 2013; 138: 787–793.

[22] Phuong DD, Trinh P Van, An N Van, et al. Effects of carbon nanotube content and annealing temperature on the hardness of CNT reinforced aluminum nanocomposites processed by the high pressure torsion technique. *J Alloys Compd* 2014; 613: 68–73.

[23] Chen B, Kondoh K, Imai H, et al. Simultaneously enhancing strength and ductility of carbon nanotube/aluminum composites by improving bonding conditions. *Scr Mater* 2016; 113: 158–162.

[24] Borkar T, Hwang J, Hwang JY, et al. Strength versus ductility in carbon nanotube reinforced nickel matrix nanocomposites. *J Mater Res* 2014; 29: 761–769.

[25] Hwang JY, Lim BK, Tiley J, et al. Interface analysis of ultra-high strength carbon nanotube/nickel composites processed by molecular level mixing. *Carbon* 2013; 57: 282–287.

[26] Suarez S, Lasserre F, Mücklich F. Mechanical properties of MWNT/Ni bulk composites: Influence of the microstructural refinement on the hardness. *Mater Sci Eng A* 2013; 587: 381–386.

[27] Suárez S, Rosenkranz A, Gachot C, et al. Enhanced tribological properties of MWCNT/Ni bulk composites – Influence of processing on friction and wear behaviour. *Carbon* 2014; 66: 164–171.

[28] Suárez S, Ramos-Moore E, Lechthaler B, et al. Grain growth analysis of multiwalled carbon nanotube-reinforced bulk Ni composites. *Carbon* 2014; 70: 173–178.

[29] Suarez S, Lasserre F, Prat O, et al. Processing and interfacial reaction evaluation in MWCNT/Ni composites. *Phys Status Solidi* 2014; 211: 1555–1561.

[30] Reinert L, Zeiger M, Suarez S, et al. Dispersion analysis of carbon nanotubes, carbon onions, and nanodiamonds for their application as reinforcement phase in nickel metal matrix composites. *RSC Adv* 2015; 5: 95149–95159.

[31] Suarez S, Lasserre F, Soldera F, et al. Microstructural thermal stability of CNT-reinforced composites processed by severe plastic deformation. *Mater Sci Eng A* 2015; 626: 122–127.

[32] Jenei P, Gubicza J, Yoon EY, et al. High temperature thermal stability of pure copper and copper-carbon nanotube composites consolidated by High Pressure Torsion. *Compos Part A Appl Sci Manuf* 2013; 51: 71–79.

[33] Jenei P, Yoon EY, Gubicza J, et al. Microstructure and hardness of copper–carbon nanotube composites consolidated by High Pressure Torsion. *Mater Sci Eng A* 2011; 528: 4690–4695.

[34] Pham VT, Bui HT, Tran BT, et al. The effect of sintering temperature on the mechanical properties of a Cu/CNT nanocomposite prepared via a powder metallurgy method. *Adv Nat Sci Nanosci Nanotechnol* 2011; 2: 015006.

[35] Chen WX, Tu JP, Wang LY, et al. Tribological application of carbon nanotubes in a metal-based composite coating and composites. *Carbon* 2003; 41: 215–222.

[36] Hwang J, Singh A, Banerjee R, et al. Processing and thermal conductivity of carbon nanotube-reinforced nickel matrix composites. *ASME Summer Heat Transf Conf* 2009; 1–4.

[37] Scharf TW, Neira A, Hwang JY, et al. Self-lubricating carbon nanotube reinforced nickel matrix composites. *J Appl Phys* 2009; 106: 013508.

[38] Singh A. RP, Hwang JY, Scharf TW, et al. Bulk nickel–carbon nanotube nanocomposites by laser deposition. *Mater Sci Technol* 2010; 26: 1393–1400.

[39] Xu C, Wei B, Ma R, et al. Fabrication of aluminum-carbon nanotube composites and their electrical properties. *Carbon* 1999; 37: 855–858.

[40] Kuzumaki T, Miyazawa K, Ichinose H, et al. Processing of carbon nanotube reinforced aluminium composite. *J Mater Res* 1998; 13: 2445–2449.

[41] Tokunaga T, Kaneko K, Horita Z. Production of aluminum-matrix carbon nanotube composite using high pressure torsion. *Mater Sci Eng A* 2008; 490: 300–304.

[42] Cho S, Kikuchi K, Miyazaki T, et al. Multiwalled carbon nanotubes as a contributing reinforcement phase for the improvement of thermal conductivity in copper matrix composites. *Scr Mater* 2010; 63: 375–378.

[43] Rajkumar K, Aravindan S. Tribological studies on microwave sintered copper–carbon nanotube composites. *Wear* 2011; 270: 613–621.

[44] Guiderdoni C, Estournès C, Peigney a., et al. The preparation of double-walled carbon nanotube/Cu composites by spark plasma sintering, and their hardness and friction properties. *Carbon* 2011; 49: 4535–4543.

[45] Sule R, Olubambi PA, Sigalas I, et al. Effect of SPS consolidation parameters on submicron Cu and Cu–CNT composites for thermal management. *Powder Technol* 2014; 258: 198–205.

[46] Yamanaka S, Gonda R, Kawasaki A, et al. Fabrication and thermal properties of carbon nanotube/nickel composite by spark plasma sintering method. *Mater Trans* 2007; 48: 2506–2512.

[47] Suárez S, Soldera F, Oliver CG, et al. Thermomechanical Behavior of Bulk Ni/MWNT Composites Produced via Powder Metallurgy. *Adv Eng Mater* 2012; 14: 499–502.

[48] Nguyen J, Holland TB, Wen H, et al. Mechanical behavior of ultrafine-grained Ni-carbon nanotube composite. *J Mater Sci* 2014; 49: 2070–2077.

[49] Rossi P, Suarez S, Soldera F, et al. Quantitative Assessment of the Reinforcement Distribution Homogeneity in CNT/Metal Composites. *Adv Eng Mater* 2015; 17: 1017–1021.

[50] Nam DH, Kim YK, Cha SI, et al. Effect of CNTs on precipitation hardening behavior of CNT/Al-Cu composites. *Carbon* 2012; 50: 4809–4814.

[51] Kim KT, Cha S Il, Hong SH. Hardness and wear resistance of carbon nanotube reinforced Cu matrix nanocomposites. *Mater Sci Eng A* 2007; 449-451: 46–50.

[52] Kim KT, Eckert J, Liu G, et al. Influence of embedded-carbon nanotubes on the thermal properties of copper matrix nanocomposites processed by molecular-level mixing. *Scr Mater* 2011; 64: 181–184.

[53] Lal M, Singhal SK, Sharma I, et al. An alternative improved method for the homogeneous dispersion of CNTs in Cu matrix for the fabrication of Cu/CNTs composites. *Appl Nanosci* 2012; 3: 29–35.

[54] Tsai P-C, Jeng Y-R. Experimental and numerical investigation into the effect of carbon nanotube buckling on the reinforcement of CNT/Cu composites. *Compos Sci Technol* 2013; 79: 28–34.

[55] Chu K, Guo H, Jia C, et al. Thermal properties of carbon nanotube-copper composites for thermal management applications. *Nanoscale Res Lett* 2010; 5: 868–874.

[56] Liu Q, Ke L, Liu F, et al. Microstructure and mechanical property of multi-walled carbon nanotubes reinforced aluminum matrix composites fabricated by friction stir processing. *Mater Des* 2013; 45: 343–348.

[57] Isaza Merino C A., Meza Meza JM, Sierra Gallego G a. A Novel Technique for Production of Metal Matrix Composites Reinforced With Carbon Nanotubes. *J Manuf Sci Eng* 2016; 138: 024501–1/5.

[58] Esawi AMK, Morsi K, Sayed A, et al. Fabrication and properties of dispersed carbon nanotube–aluminum composites. *Mater Sci Eng A* 2009; 508: 167–173.

[59] Choi HJ, Min BH, Shin JH, et al. Strengthening in nanostructured 2024 aluminum alloy and its composites containing carbon nanotubes. *Compos Part A Appl Sci Manuf* 2011; 42: 1438–1444.

[60] Choi HJ, Shin JH, Bae DH. Grain size effect on the strengthening behavior of aluminum-based composites containing multi-walled carbon nanotubes. *Compos Sci Technol* 2011; 71: 1699–1705.

[61] Asgharzadeh H, Joo S-H, Kim HS. Consolidation of Carbon Nanotube Reinforced Aluminum Matrix Composites by High-Pressure Torsion. *Metall Mater Trans A* 2014; 45: 4129–4137.

[62] Carvalho O, Miranda G, Soares D, et al. Carbon nanotube dispersion in aluminum matrix composites — Quantification and influence on strength. *Mech Adv Mater Struct* 2016; 23: 66–73.

[63] Tu JP, Yang YZ, Wang LY, et al. Tribological properties of carbon-nanotube-reinforced copper composites. *Tribol Lett* 2001; 10: 225–228.

[64] Kim KT, Cha S Il, Hong SH. Microstructures and tensile behavior of carbon nanotube reinforced Cu matrix nanocomposites. *Mater Sci Eng A* 2007; 448-451: 46–50.

[65] Li H, Misra A, Zhu Y, et al. Processing and characterization of nanostructured Cu-carbon nanotube composites. *Mater Sci Eng A* 2009; 523: 60–64.

[66] Li H, Misra A, Horita Z, et al. Strong and ductile nanostructured Cu-carbon nanotube composite. *Appl Phys Lett* 2009; 95: 25–27.

[67] Uddin SM, Mahmud T, Wolf C, et al. Effect of size and shape of metal particles to improve hardness and electrical properties of carbon nanotube reinforced copper and copper alloy composites. *Compos Sci Technol* 2010; 70: 2253–2257.

[68] Chu K, Jia C, Li W, et al. Mechanical and electrical properties of carbon-nanotube-reinforced Cu-Ti alloy matrix composites. *Phys Status Solidi* 2013; 210: 594–599.

[69] Shukla a. K, Nayan N, Murty SVSN, et al. Processing of copper-carbon nanotube composites by vacuum hot pressing technique. *Mater Sci Eng A* 2013; 560: 365–371.

[70] Chu K, Jia C, Jiang L, et al. Improvement of interface and mechanical properties in carbon nanotube reinforced Cu–Cr matrix composites. *Mater Des* 2013; 45: 407–411.

[71] Yoon EY, Lee DJ, Park B, et al. Grain refinement and tensile strength of carbon nanotube-reinforced Cu matrix nanocomposites processed by high-pressure torsion. *Met Mater Int* 2013; 19: 927–932.

[72] Barzegar Vishlaghi M, Ataie A. Investigation on solid solubility and physical properties of Cu–Fe/CNT nano-composite prepared via mechanical alloying route. *Powder Technol* 2014; 268: 102–109.

[73] Akbarpour MR, Farvizi M, Lee DJ, et al. Effect of high-pressure torsion on the microstructure and strengthening mechanisms of hot-consolidated Cu–CNT nanocomposite. *Mater Sci Eng A* 2015; 638: 289–295.

[74] Varo T, Canakci A. Effect of the CNT Content on Microstructure, Physical and Mechanical Properties of Cu-Based Electrical Contact Materials Produced by Flake Powder Metallurgy. *Arab J Sci Eng* 2015; 40: 2711–2720.

[75] Imai H, Kondoh K, Li S, et al. Microstructural and Electrical Properties of Copper - Titanium Alloy Dispersed with Carbon Nanotubes via Powder Metallurgy Process. *Mater Trans* 2014; 55: 522–527.

[76] Hwang JY, Neira A, Scharf TW, et al. Laser-deposited carbon nanotube reinforced nickel matrix composites. *Scr Mater* 2008; 59: 487–490.

[77] Choi H, Shin J, Bae D. The effect of milling conditions on microstructures and mechanical properties of Al/MWCNT composites. *Compos Part A Appl Sci Manuf* 2012; 43: 1061–1072.

[78] Suárez S, Ramos-Moore E, Mücklich F. A high temperature X-ray diffraction study of the influence of MWCNTs on the thermal expansion of MWCNT/Ni composites. *Carbon* 2013; 51: 404–409.

[79] Kwon H, Park DH, Silvain JF, et al. Investigation of carbon nanotube reinforced aluminum matrix composite materials. *Compos Sci Technol* 2010; 70: 546–550.

[80] Sairam K, Sonber JK, Murthy TSRC, et al. Influence of spark plasma sintering parameters on densification and mechanical properties of boron carbide. *Int J Refract Met Hard Mater* 2014; 42: 185–192.

[81] Suk-Joong L.Kang. Sintering Densification, Grain Growth, and Microstructure. *Elsevier Butterworth-Heinemann Linacre House* 2005; pp. 9–18.

[82] Cumings J, Zettl A. Electrical and Mechanical Properties of Nanotubes Determined Using In-situ TEM Probes. In: Rotkin S, Subramoney S (eds) *Applied Physics of Carbon Nanotubes*. Springer, 2005, p. 349.

[83] Huang Q, Gao L, Liu Y. Sintering and thermal properties of multiwalled carbon nanotube – BaTiO3 composites. *J Mater Chem* 2005; 1995–2001.

[84] Lahiri D, Singh V, Keshri AK, et al. Carbon nanotube toughened hydroxyapatite by spark plasma sintering: Microstructural evolution and multiscale tribological properties. *Carbon* 2010; 48: 3103–3120.

[85] Zhan G-D, Kuntz JD, Wan J, et al. Single-wall carbon nanotubes as attractive toughening agents in alumina-based nanocomposites. *Nat Mater* 2003; 2: 38–42.

[86] Hu N, Jia B, Arai M, et al. Prediction of thermal expansion properties of carbon nanotubes using molecular dynamics simulations. *Comput Mater Sci* 2012; 54: 249–254.

[87] Kwon Y-K, Berber S, Tománek D. Thermal Contraction of Carbon Fullerenes and Nanotubes. *Phys Rev Lett* 2004; 92: 015901.

[88] Berber S, Kwon Y, Tomanek D. Unusually high thermal conductivity of carbon nanotubes. *Phys Rev Lett* 2000; 84: 4613–6.

[89] Kim P, Shi L, Majumdar A., et al. Thermal Transport Measurements of Individual Multiwalled Nanotubes. *Phys Rev Lett* 2001; 87: 19–22.

[90] Poncharal P, Berger C, Yi Y, et al. Room temperature ballistic conduction in carbon nanotubes. *J Phys Chem B* 2002; 106: 12104–12118.

[91] Poncharal P, Frank S, Wang ZL, et al. Conductance quantization in multiwalled carbon nanotubes. *Eur Phys J D* 1999; 9: 77–79.

[92] Hone J, Llaguno MC, Nemes NM, et al. Electrical and thermal transport properties of magnetically aligned single wall carbon nanotube films. *Appl Phys Lett* 2000; 77: 666–668.

[93] Bakshi SR, Batista RG, Agarwal A. Quantification of carbon nanotube distribution and property correlation in nanocomposites. *Compos Part A Appl Sci Manuf* 2009; 40: 1311–1318.

[94] Bakshi SR, Lahiri D, Agarwal A. Carbon nanotube reinforced metal matrix composites - a review. *Int Mater Rev* 2010; 55: 41–64.

[95] Agarwal A, Bakshi SR, Lahiri D. *Carbon Nanotubes reinforced Metal Matrix Composites.* 1st ed. Boca Raton, FL, USA: CRC Press, 2011.

[96] Okamoto H. Phase Diagram Updates: Section III: Al-C (Aluminum-Carbon). *J Phase Equilibria* 1992; 13: 97–98.

[97] Lee J, Novikov N (eds). *Innovative superhard materials and sustainable coatings for advanced manufacturing*. Dordrecht, The Netherlands: Springer - NATO Science Series, 2005.

[98] Foster LM, Long G, Hunter MS. Reactions Between Aluminum Oxide and Carbon The Al2O3—Al4C3 Phase Diagram. *J Am Ceram Soc* 1956; 39: 1–11.

[99] Yamada Y, Castleman Jr AW. Gas-phase copper carbide clusters. *Chem Phys Lett* 1993; 204: 133–138.

[100] Park M, Kim BH, Kim S, et al. Improved binding between copper and carbon nanotubes in a composite using oxygen-containing functional groups. *Carbon* 2011; 49: 811–818.

[101] Singleton M, Nash P. The C-Ni (Carbon-Nickel) system. *Bull Alloy Phase Diagrams* 1989; 10: 121–126.

[102] Hwang JY, Singh ARP, Chaudhari M, et al. Templated Growth of Hexagonal Nickel Carbide Nanocrystals on Vertically Aligned Carbon Nanotubes. *J Phys Chem C* 2010; 114: 10424–10429.